Contents

Preface

When this book first appeared it was designed as an intro-
ductory book for those with little or no knowledge of microbiology. In the intervening fifteen years not only has the
understanding of micro-organisms advanced considerably
but we have witnessed many changes in secondary education
such that microbiology is now a component of most school
curricula in biology. This revision has attempted to be true to
the aim of the original author whilst strengthening the
breadth of coverage through revision of the component
chapters. In many cases this has involved a substantial re-
writing of the material with the introduction of more recently
developed concepts to give the reader a more balanced
impression of microbiology in the 1980s. We have also
recognized the growing importance of the biotechnological
applications of micro-organisms by devoting a single chapter
to this important aspect of microbiology. Thus, by reading
this text it is hoped that the reader will gain an overview of the
current status of the many diverse and interesting aspects of
modern microbiology.

We would like to thank our secretaries, Sylvia McKenzie
and Gill Taylor, for their perseverence in the eternal process
of redrafting the component parts of the book. Our thanks
are also extended to those who have given original illus-
trations for reproduction in this book. Finally, I would like to
thank my co-revisionists for expediting the modernization of
this text.

Ian R. Booth

1: Introduction

What is microbiology?

Microbiology is the study of micro-organisms, that is organisms which are of microscopic dimensions. Because the human eye cannot resolve any object smaller than 1/10 mm in diameter and most micro-organisms are only a few thousandths of a millimeter in size, they can only be seen with the aid of a microscope. As a direct consequence of the 'invisibility' of microbes to the naked eye and the need for specialized techniques to study them, microbiology was the last of the three major divisions in biology to develop. However, if it was late to arrive, it is a discipline which has been quick to develop.

It is now usual to include five major groups as micro-organisms; the subdivisions of virology, bacteriology, mycology, phycology and protozoology (the studies of viruses, bacteria, fungi, algae and protozoa respectively).

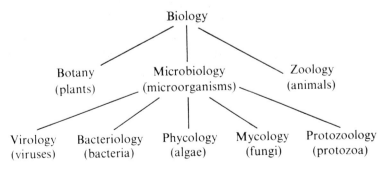

There have been continued attempts to include all organisms under the disciplines of either botany or zoology; bacteria, algae and fungi have in the past been considered to be part of the Plant Kingdom while protozoa have been included in the Animal Kingdom. This type of classification cannot be supported from a taxonomic point of view.

If we group all living things according to the similarities of certain nucleic acids which are common to all organisms (the ribosomal RNAs), it is found that the most distantly related organisms (ones which diverged earliest during evolution) fall naturally into three groups. Accordingly, all animals, plants, algae, fungi and protozoa belong to just one group, the 'eukaryotes', which are distinguishable from the 'eubacteria' or true bacteria and the archaebacteria or primitive bacteria. Since microbiology encompasses the study of groups of organisms in all three of these kingdoms it may be argued that it covers a greater biological diversity than its sister divisions rather than vice versa.

The diversity of life within the microbiological world may appear less obvious to many than that of types of mammals or flowering plants. However, the striking diversity of microbial organisms lies not in their range of morphologies but rather in their ecological and physiological specializations. Bacteria have existed for at least three 3×10^9 of the earth's 4.5×10^9 year life span and in this time have evolved to cope with and flourish in almost every niche, no matter how inhospitable.

There are microbes which are adapted for life in the coldest oceans and in hot water springs where temperatures approach boiling point. Others are capable of growing in saturated salts, at high pressures, in acid at pH 0.2 or alkali at pH 12.5. The most radiation resistant organism known is a bacterium—*Deinococcus radiodurans*. The versatility of microbes to cope with extremes of environment extends also to the range of substances from which various micro-organisms can obtain energy and cell nutrients. Micro-organisms are the only group of organisms capable of fixing atmospheric nitrogen into utilizable organic compounds, or of growth in the total absence of oxygen. The terrestrial ecosystem depends on the activities of bacteria and fungi to dispose of the organic detritus that would otherwise accumulate and consume the land. To this end there are microbes that are capable of biodegrading every natural organic compound and all but a few of the most recalcitrant man-made organic materials such as certain plastics, dioxin, polyfluorocarbons etc.

Micro-organisms, as we shall see, also form a range of associations with other microbes and with higher plants and animals. They can be pathogens, parasites, symbionts, commensals and saprophytes, and thus their ecological influences infiltrate into all the trophic levels of life and the gamut of possible ecosystems. The microbiological world

2

may be hidden from sight but it is a microcosm whose activities are of central importance to the structure of the biosphere.

Microbes and man

Self-centredly we tend to evaluate the importance of things by their impact on ourselves. Despite their small size, microbes are certainly of immense importance to man; they cause disease, provide us with various foods and medicines, and dispose of our wastes. In a real sense they are responsible for the very air we breathe since free molecular oxygen was completely absent from the pre-biotic atmosphere and has accumulated only as a product of the metabolism of primeval photosynthetic bacteria.

Mankind has made use of micro-organisms, or their bio-chemical activities, since long before he even knew of their existence. We know that in 6 000 BC the ancient Babylonians and Sumerians were brewing beer much as we do today and that the Egyptians were baking leaven bread 2 000 years later. Despite the antiquity of these microbiological practices, the first documentations of the structure of micro-organisms did not occur until the advent of the first microscopes in the seventeenth century. Although the spontaneous generation of mice from old rags and of maggots from meat had been disproved by this time, it was not until last century that Louis Pasteur showed that microbes were not produced *de novo* from muds or decaying organic matter.

If Pasteur was the founder of industrial microbiology, Robert Koch was the forefather of medical microbiology. Koch, a German doctor, showed in 1876 that the causative agent for bovine anthrax was a bacterium, *Bacillus anthracis*. He isolated the bacillus from diseased cattle, cultured it on nutrient jelly and showed that the cultured micro-organisms would cause the symptoms of anthrax when inoculated into healthy cattle. This remarkable work remains a milestone in microbiology and revolutionized mankind's perception of the nature of disease.

We live in a time when microbiology has come of age. Industrial microbiologists produce microbial products on a huge scale—14 000 tons of penicillin, 300 million tons of the flavour enhancer MSG (monosodium glutamate), and 1 430 million tons of vinegar are manufactured per annum. We use microbes to make beer, wine, cheese, yogurt, sauerkraut, soya sauce, antibiotics, pesticides, gels and many other products. Microbiological reactions are used to process sewage, transform the chemical structures of drugs, clean clothes (bacterial enzymes are used in biological detergents) and even to extract precious metals such as copper and uranium from their mineral ores. Within the last decade new technologies have been developed, such as gene cloning, which will use microbes as factory cells for the synthesis of valuable pharmaceutical products such as human insulin, hormones, antiviral drugs and vaccines. When somatostatin, a growth hormone secreted by the brain, was first purified it took 500 000 sheep brains to produce a five milligram quantity. In 1977, genetic engineers produced an equal quantity of somatostatin from two gallons of bacterial culture at a cost of less than £5.

Despite the dramatic advances in medical microbiology since the time of Robert Koch, micro-organisms will continue to be a major problem in medicine and in diseases of plants. The outbreak of influenza in 1918–1919 claimed more than 20 million lives—far more than were killed during the World War which was then coming to an end. Each year fungal, bacterial and viral diseases of plants cause some 3.3 billion dollars loss in the world's food crops. There is still no cure for the common cold or any of the more serious viral diseases of man and even eminently curable bacterial diseases such as tuberculosis and leprosy still affect millions of people in the underdeveloped nations of the world. We continue the work towards producing an effective vaccine against the protozoan malarial parasite which kills over a million people each year. Also, we will have to continue to meet the challenge of diagnosis and treatment of 'new' diseases such as Legionnaires' disease and AIDS. It has also become evident that viruses are involved in the aetiology of some forms of cancer. It would, however, be grossly misleading to create the impression that micro-organisms such as bacteria are by their very nature pathogenic. A normal healthy human body harbours on its surface and within its alimentary canal ten times as many microbial cells than it has cells of its own kind. Many of these are of positive benefit to the digestive process and the rest are mostly harmless passengers which we never notice.

The fascination of microbes to microbiologists lies not only in their utility and pathogenic activities but also on the powerful insights we glean from studies of microbes about the ways in which living things of all types grow and multiply. It may seem strange to some that we know more about the detailed biology of a bacterium (*Escherichia coli*) which lives in the human intestine than any other organism. Indeed, this bacterium has been used as the organism of choice during the past few decades in research concerned with the formulation

4

of general biological principles and particularly of those at a molecular level. As Francis Bacon noted, the nature of things is commonly better perceived in small than in great. Our relationships with the microbes is and will continue to be dichotomous—they are our deadliest adversaries but also our closest allies.

Methods in microbiology

A subject can only develop according to the techniques available. This may sound platitudinous but a study of the history of microbiology demonstrates the prime importance of suitable methodology. Three techniques in particular had to be perfected before the science of microbiology could evolve beyond a primitive visionary state.

1 *Microscopy*. Since microbiology is mainly concerned with the study of living organisms of microscopic dimensions, its development depended for its initiation entirely upon the refinement of the microscope.

2 *Sterilization methods*. Media to be used for growth of a particular micro-organism had to be freed from all other living organisms; in other words, sterilization methods had to be developed.

3 *Pure culture methods*. Once it was possible to obtain sterile growth media, it became practicable to introduce methods to separate different micro-organisms from each other and to maintain them in pure culture. Their individual characteristics could then be studied. Let us consider each of these critical developments in turn.

The microscope

Prior to the seventeenth century there had been various reports of the existence of invisible living creatures but, before the development of suitable means of magnifying them, no proof was obtainable. To Anthonie van Leeuwenhoek, a Dutch merchant and amateur scientist living in Delft, belongs the honour of providing the first accurate report of the occurrence of bacteria. Leeuwenhoek employed his spare time in pursuing his hobby of making lenses which he used to build magnifying glasses of high resolving power. These single-lens microscopes were of the simplest possible design (see Fig. 1.1) but were still capable of magnifying an object by about 200-fold. As a result of his exceptionally painstaking care in the building and use of his microscopes, Leeuwenhoek was able to make descriptions of many micro-organisms including some which were almost certainly bacteria.

5

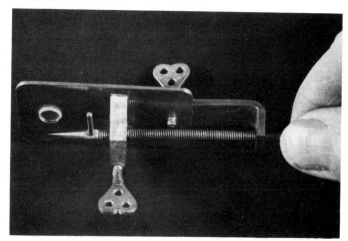

Fig. 1.1. The type of microscope used by Leeuwenhoek. The object is placed at the end of a spike attached to a screw and is viewed through a small lens.

Anybody who has tried to use a reproduction of one of his instruments will soon realize that his particular genius fulfils the criterion of an infinite capacity to take pains. Using a racey style that would nowadays be the subject of multitudinous editorial transformation, he communicated the results of his work in the form of letters to the recently founded Royal Society of London. An example is given in this excerpt of a letter of 1684 which gives the first description of bacteria:

> 'Though my teeth are kept usually very clean, nevertheless when I view them in a magnifying glass, I find growing between them a little white matter . . . I took some of this flower and mixed it with pure rain water wherein were no animals . . . and to my great surprise perceived that the aforesaid matter contained many small living animals, which moved themselves very extravagantly. The biggest sort had the shape of A, their

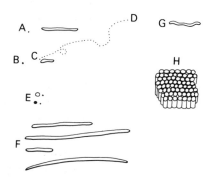

Fig. 1.2. Leeuwenhoek's famous drawing of micro-organisms published in 1604. Different shaped organisms can be seen, some of which were most certainly bacteria.

6

motion was strong and nimble, and they darted themselves through the water as a Jack or Pike does through water. . . . The second sort had the shape of B. These spun about like a Top. . . . In the third sort I could not well distinguish the figure, for sometimes it seemed to be an oval and other times a circle. These were so small that they seemed no bigger than E'.

These observations of Leeuwenhoek were followed by a period of nearly two centuries during which there was little further descriptive work on the smaller micro-organisms because nobody was capable of designing a microscope which had a sufficient resolving power but which could, at the same time, be used by a worker of average dexterity and patience. Further progress depended upon the development of a compound microscope with an eyepiece and objective lens allowing an increase in the magnification obtainable and greater ease of manipulation. It is true that Robert Hooke had used compound microscopes in the seventeenth century but they were incapable of the performance given by Leeuwenhoek's single-lens microscopes. The reason for this lay in defects such as chromatic and spherical aberration inherent in their basic design. During the eighteenth century these defects were gradually overcome by the following refinements:

1   Corrected complex eyepiece and objective lenses.
2   A condenser to focus light on the object.
3   A thin glass coverslip to place over a liquid drop on a glass slide so that objects within the liquid could be viewed in a flat plane.
4   The oil-immersion lens. The resolving power of a microscope can be increased by using a material lens of higher refractivity than air between the objective. The material most commonly used is a special immersion oil.

In conjunction with these developments in microscopic design, staining methods were perfected so as to allow a simple classification of micro-organisms based on morphological grounds. However, the theoretical limit of resolution of the light microscope is about 0.2 μm and clearly gives little hope of seeing much internal structural detail in a typical bacterium of 0.5–1.0 μm diameter.

Just as there was a quiescent period of microbial cytology between the work of Leeuwenhoek and the sophistication of the compound microscope, so there was little real progress between the latter half of the nineteenth century and the development of the electron microscope in the 1940s. The small wavelength of an electron beam allows a theoretical resolving power down to 0.01 nm and for the first time, viruses could be demonstrated as physical entities. Practical difficulties in instrument design such as the development of magnetic lenses have prevented resolutions as low as this but it is still possible to see the larger molecules that make up the architecture of a cell. In practice only thin objects can be viewed with any real hope of obtaining good definition of internal structure and so methods had to be developed in which cells are fixed, dehydrated, embedded in plastic and sectioned to give a preparation about 100 nm thick (i.e. about 10 sections to a bacterial cell). Some increase in contrast can be obtained by using electron-dense stains like osmic acid, permanganate, or uranium salts. Another problem in electron microscopy is the possibility of artefacts caused by fixation, drying and embedding; this difficulty can be partly obviated by the use of freeze-etching in which a carbon replica is made of a frozen surface in a cell (Fig. 2.8). In spite of the many difficulties involved in its use and particularly in the interpretation of results, the electron microscope has opened up a new world to microbial cytologists.

Methods of sterilization

Sterilization involves the complete destruction or removal of all living organisms from the object being sterilized. The development of methods for sterilization was very largely a happy consequence of the controversy over spontaneous generation culminating in the work of Pasteur.

Experiments designed to prove or to disprove spontaneous generation depended upon two general principles:
1 The complete sterilization of a suitable growth medium so that no living organisms exist at the start of the experiment.
2 The design of the vessel so that it is impossible for micro-organisms to enter from the outside. This was necessary following the realization of the existence of micro-organisms floating around in the air. For example, even 'fresh' air may contain one particle carrying a micro-organism per cubic foot while the figure may be a hundred to a thousand times greater in a crowded room.

Provided these principles are rigidly adhered to and provided the conditions are otherwise suitable for microbial multiplication, any growth occurring must be the result of spontaneous generation. Clearly the key question was how good the methods were for attaining and maintaining sterility. Such was the emotional fervour aroused by a controversy which involved the very nature of life that many important scientists became involved. The technical 'fall-out' was the development of sterilization methods. Let us consider the two principles in greater detail:

1 The attainment of sterility. The usual method depended upon heat treatment, which was known to be inimical to most forms of life. However, it was soon realized that micro-organisms vary widely in their resistance to heating (Fig. 1.4) and sterilization clearly must be gauged to the most resistant form. In general, bacteria required higher temperatures than larger forms and some micro-organisms can produce specialized heat-stable structures called spores (p. 27). Boiling at normal pressure was insufficient to kill these spores and so autoclaves were designed to increase the pressure, and, thereby, the temperature.

2 The maintenance of sterility. In experiments claiming to show spontaneous generation, a cork was often used to prevent the entry of contaminants from outside. Unfortunately, this method was ineffective in practice since micro-organisms could enter round the side of the cork as the vessels cooled after sterilization. Although a flask could be hermetically sealed, this led to the objection that oxygen, a substance known to be essential for many forms of life, could no longer enter the vessel. It was necessary, therefore, to include some sort of filter to prevent the entry of micro-organisms but not of air. This led to the development of the cotton wool plug which was soon adopted universally by microbiologists. However, one of the simplest and most elegant means of preventing the entry of micro-organisms can be seen in Pasteur's swan-necked flask (see Fig. 1.3). This worked on the basis that organisms in the atmosphere entering the open end of the tube would, in their slow passage through the convolutions, be deposited by the pull of gravity. Pasteur showed that such flasks, although left open, remained sterile indefinitely. By such simple means, he finally disproved the idea of spontaneous generation, a result aided by his skill as an expositor of his own work and by an almost evangelical zeal.

However, although the brilliance and the force of Pasteur's personality were able to catalyse the early adolescence of microbiology, he later acted as an inhibitor of progress since it was often felt that 'surely Pasteur can't be wrong'. For example, he firmly believed that metabolic

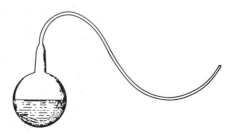

Fig. 1.3. The Swan-necked flask used by Pasteur to demonstrate the absence of spontaneous generation.

processes such as fermentation were an essential property of the 'living force' and could not occur in the absence of living cells, and for a long time this view was generally supported. In science, as elsewhere, an inhibition of progress is the price we may have to pay for an innovative genius.

By the end of the nineteenth century most of the methods currently used for sterilization had been developed. These are briefly summarized below.

Heat

If the percentage of survivors of a micro-organism is plotted against time a logarithmic relationship is found. The slope of the line varies from organism to organism but in the choice of a general sterilization method we must use a time and temperature that will kill all organisms including heat-resistant spores.

This is illustrated in Fig. 1.4 where it can be seen that if a time of 30 minutes is used for sterilization, micro-organism A will require a temperature of only 50° while B needs 60–70° and the bacterial spore suspension C, 120°. The methods generally adopted are as follows:

1   *Wet heat in an autoclave.* The usual method is a time of 30 minutes at a pressure of 1.05 kg/cm² which will give a temperature of 121°. This is the best method if it is practicable.

2   *Tyndallization.* A course of three periods of boiling at 100° for 30 minutes at daily intervals. The spores remaining

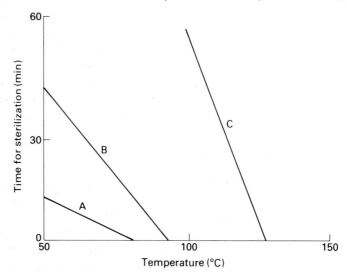

Fig. 1.4. The death curves resulting from heat treatment of two suspensions of vegetative cells of micro-organisms A and B and a spore suspension from micro-organism C.

at the end of the first stage will germinate as the temperature is lowered; this results in a loss of heat-resistance so that the vegetative cells are killed at the second or third boiling. This process is occasionally used for materials such as sugar media likely to be affected by the higher temperatures used in an autoclave.

3   *Dry heat*. Water acts as a catalyst in the killing of micro-organisms by heat and in its absence, as in a dry oven, a temperature of 160° for two hours is usually required.

4   *Pasteurization*. This is *not* a method of sterilization since it involves treatment for an insufficient time for complete killing, e.g. to about 60°C for 30 minutes. This treatment will kill most disease-producing bacteria present in natural products such as milk without affecting the flavour or consistency.

Filtration

The liquid or gas to be sterilized is passed through a filter with a porosity sufficient to remove any micro-organisms in suspension. The use of cotton wool for gases has already been mentioned and for liquids a variety of filters are available, made of materials such as cellulose nitrate (Millipore filters). Filtration is obviously the method of choice for the sterilization of liquids containing heat-labile components.

Radiation

Certain regions of the radiation spectrum are lethal to micro-organisms. Ultraviolet light is especially effective and is valuable in sterilizing air. However, it penetrates poorly and for the interior of solid objects it is necessary to use ionizing radiations such as gamma rays or X-rays from a source such as radioactive cobalt. Ionizing radiation is often used to sterilize plastics and other heat-labile materials.

Chemicals

Many chemicals are lethal to micro-organisms in general but they may require a long and impracticable exposure for complete sterilization and are often difficult to remove subsequently. Hypochlorite solutions and phenolic derivatives are used as general laboratory disinfectants, as is gaseous ethylene oxide for more specialized purposes. However, it must be realized that the chemicals often used as disinfectants rarely cause sterilization under the conditions employed.

Pure culture methods

As soon as microbial growth media could be sterilized effectively it became theoretically possible to isolate pure cultures. However, although Pasteur had shown that different fermentation and putrefaction processes were associated with the growth of morphologically different microbes (e.g. yeast with alcohol fermentation, a variety of shapes of bacteria with lactic, acetic and butyric fermentations), some scientists held that all micro-organisms, and particularly all bacteria, were variants of a single basic type— the concept of *pleomorphism*. Others believed in *monomorphism*—i.e. in a large number of different organisms which should all be obtainable in pure culture. Could they? Ultimately, any method must depend upon the introduction of a single microbial cell into a sterile growth medium in a suitable vessel. Unfortunately, the small size of most micro-organisms made mechanical separation of single cells impossible. It is true that the more recent development of the micromanipulator has made this feat possible, but it is a difficult and specialized instrument for general use.

Other methods had to be discovered which had the effect of diluting a sample so that single cells were obtained which could grow to produce a pure culture. The first methods depended upon a dilution of the culture until an aliquot was likely to contain a single cell as judged from an initial count. However, such methods are tedious, unreliable and can only

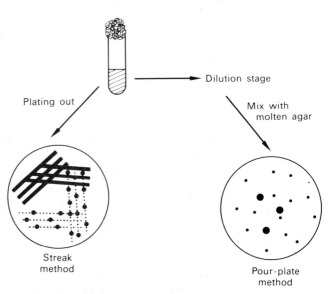

Fig. 1.5. Methods for obtaining a pure culture of a micro-organism. The original liquid culture in the example contains two organisms, one producing large colonies and the other small colonies.

be employed for the dominant organisms. They are now rarely used and instead microbiologists began to study the possibility of diluting on a solid surface. The sample is placed at one point on a sterile solid growth medium and then, using a sterile needle, the sample, or inoculum as it is called, is drawn several times over the surface. Each streak represents a dilution process and eventually single cells are obtained along the streak. Each of these, on incubation, grows up into a separate colony which can be used as the source of a pure culture. This streak method for the isolation of pure cultures was pioneered by Robert Koch who realized that the most effective surface to use was a solidified form of a normal growth medium. At first he employed gelatin as a solidifying agent but it suffered from liquefaction at temperatures above 28°C and, being a protein, it can be hydrolysed by some micro-organisms. A nearly ideal substance was found in the polysaccharide agar–agar (usually simply called agar) occurring in some seaweeds; an agar gel melts at about 100°C and will not solidify again until the temperature falls to about 40°C. Furthermore, agar is attacked by very few micro-organisms. Its unique properties allowed an alternative to the streak method—the pour plate method. A diluted sample of the micro-organism is mixed with a previously melted agar growth medium at a temperature just above the solidifying point. The mixture is poured into a suitable vessel and is incubated. Each cell produces a colony within the agar. Techniques applied to a whole range of microbes occurring in natural environments. However, if a nutrient medium suitable for the growth of most micro-organisms is employed, only the predominant organisms are likely to be isolated. Take, for example, the micro-organisms inhabiting our mouths, a typical sample may contain the following groups of organisms in numbers per ml of saliva: yeasts 100; lactobacilli 2×10^4; staphylococci 5×10^5; diptheroids 3×10^6; streptococci 5×10^7; anaerobic micrococci 5×10^7; anaerobic diplococci 1×10^8. If we are to isolate some of the rarer organisms in pure culture we must choose environmental conditions which favour the growth of the organism we require compared with that of those we do not. This is done either by a previous enrichment stage, or by choosing appropriate media and conditions for the dilution and growth stage. Let us consider some of the selection methods used in practice:

1 Growth at a high temperature (e.g. 50°) above the maximum for most micro-organisms will select by 'heat-loving' or thermophilic species (see p. 144). Alternatively, if

this sample is pasteurized (p. 11) before growth, in general sporing species will be isolated.

2 Growth in the absence of a source of nitrogen in the growth medium but in the presence of air will select those organisms capable of utilizing atmospheric nitrogen—the nitrogen-fixing micro-organisms (p. 68).

3 Growth in the absence of oxygen will select those organisms capable of anaerobic growth (p. 146).

4 Micro-organisms growing in mammalian intestines have to be able to withstand the presence of bile salts which inhibit the growth of most other microbes. If a medium is made up containing suitable nutrients together with bile salts, intestinal bacteria are favoured; this property is the basis of methods used for testing water supplies for faecal contamination (p. 126).

These are a few examples of the environmental conditions that can be devised to increase the percentage of the desired micro-organisms from a mixed inoculum so that single colonies can be isolated. Using these and similar methods, microbiologists were able to show the vast range of micro-organisms occurring in nature.

The development of the techniques described above allowed micro-organisms to be isolated and purified. The early stages of microbiology were then concerned with the discovery of the structure and the biochemical activity of the organisms. Ultimately, this led to a consideration of the mechanisms of heredity and of control of expression of particular properties. The last 10–15 years have seen a gradual move away from considerations of the isolated organism and there is now great interest in mixed populations of cells, and the roles of such consortia in the environment. Finally, the latest phase is the development of micro-organisms as biochemical reactors, e.g. as 'hosts' for foreign gene expression and as the producers of single enzymes or groups of enzymes concerned with specific reactions. These recent developments build upon the experience of earlier microbiologists and so an understanding of the core of microbiology is essential. The following chapters seek to describe and explain the basic aspects of structure, function and activity of micro-organisms in isolation and in populations.

2 : The Structure of Micro-organisms

The introduction of the light microscope allowed micro-biologists to determine the overall shape of cells, and in the case of multicellular organisms, their arrangement. Further-more, some internal structure can be seen by the use of the phase-contrast or interference microscopy or by staining methods with the bright-field or fluorescence microscopy. Most of the stains used, such as basic dyes, are relatively non-specific and give little idea of the chemical nature or function of a cellular component. Some are more specific; fat-soluble dyes such as Sudan Black stain lipids, iodine colours starch granules, and acridine orange fluoresces with nucleic acids. However, in small micro-organisms such as bacteria, the size of the cell (say about 0.5 μm in diameter) is only slightly greater than the theoretical limit of resolution of the light microscope (about 0.2 μm). The study of the ultra-structure of cells has depended upon the introduction of the transmission and stereoscan electron microscopes (although the possibility of artefacts produced in the preparation of specimens must always be borne in mind). Furthermore, as in all cytological investigations, it is very easy to see what one is looking for by selection of appropriate microscopic fields. Consider the well-known poem of Hilaire Belloc:

> The microbe is so very small
> You cannot make him out at all,
> But many sanguine people hope
> To see him through a microscope.
> His jointed tongue that lies beneath
> A hundred curious rows of teeth,
> His seven tufted tails with lots
> Of lovely pink and purple spots
> On each of which a pattern stands

Composed of forty separate bands;
His eyebrows of a tender green
All these have never yet been seen
But scientists who ought to know
Assure us that they must be so.
Oh, let us never, never doubt
What nobody is sure about

Although Belloc's visions leave something to be desired as an exposition of microbial structure, the final 'punch' lines are particularly appropriate to the cytologist who needs considerable integrity and perhaps not too much imagination.

One of the main problems in electron microscopy is in the interpretation of the chemical nature and function of the structures seen. Unfortunately, there are few specific stains available and although there may be potential in the use of specific antibodies labelled with electron dense components like ferritin or colloidal gold, there are technical difficulties in their use. Consequently, the primary method used for relating structure to function is to break open the cell and to fractionate the components which are then subjected to biochemical analysis. Further problems arise regarding the difficulty of rupturing a cell without denaturing the more sensitive components, particularly in bacteria with their tough walls and small size. However, methods of breakdown have gradually been developed and most of the major cell structures can be isolated in a reasonably pure state in which they perform their functions normally. Consider two examples of cell breakage in bacteria (Fig. 2.1).

1   Violent shaking of a bacterial suspension with glass beads causes rupture of the cell walls and liberation of the cytoplasmic contents. The walls can be concentrated by centrifugation and treated with appropriate solvents and enzymes to remove contaminating components. The walls isolated in this way retain the shape of the original cell. In a few bacterial species the wall can also be solubilized completely by the enzyme lysozyme, a component of numerous animal secretions and fluids.

2   Exposure of a sensitive bacterial species to lysozyme normally causes complete cell lysis. However, if the treatment is done in the presence of an osmotic stabilizer such as an isotonic sucrose solution, the cell assumes a spherical shape as the wall is solubilized.

The product is called a *protoplast* and electron microscopy of thin sections shows that it is bounded by a structure called the cytoplasmic membrane. This simple experiment suggests the following:

1 The wall determines cell shape since the protoplast is spherical irrespective of the original shape.

2 The wall is responsible for the mechanical strength of the bacterial cell since a protoplast is very prone to mechanical or osmotic lysis while the cell is not.

3 The cytoplasmic membrane and not the cell wall is responsible for the selective permeability of the cell surface; this remains unchanged in a protoplast.

Further experiments show that the protoplast retains most of the enzyme systems of the cell and is capable of regeneration and growth while the wall is not.

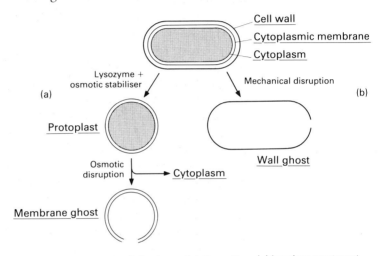

Fig. 2.1. Two methods for bacterial disruption: (a) involves treatment with lysozyme in the presence of an osmotic stabilizer followed by controlled osmotic lysis, and (b) involves mechanical disruption with glass beads.

If a protoplast is subject to controlled osmotic lysis by a gradual reduction in the external osmotic pressure, the cytoplasmic membrane ruptures, liberating the cytoplasm and its contents. Differential centrifugation can then be used to isolate the membranes and the components of the cytoplasm which are unaffected by the mild lytic procedure.

Using methods similar to those outlined above, the structure and function of the components of a few apparently typical cells have been obtained. It has become clear that there are two basic types of cell—*prokaryotic* or *eukaryotic*. Prokaryotic cells are restricted to micro-organisms (bacteria, including blue-green bacteria), while eukaryotic cells occur in micro-organisms (fungi, protozoa and algae) and in animals and plants. In addition, there are the viruses which have a much simpler non-cellular structure and will be dealt with in Chapter 6.

The prokaryotes and eukaryotes share the common basic unit of life, the cell. All cells have the following essential components:

(a) DNA, as genetic information for replication.
(b) RNA, for protein synthesis.
(c) Enzymes, for catalysis.

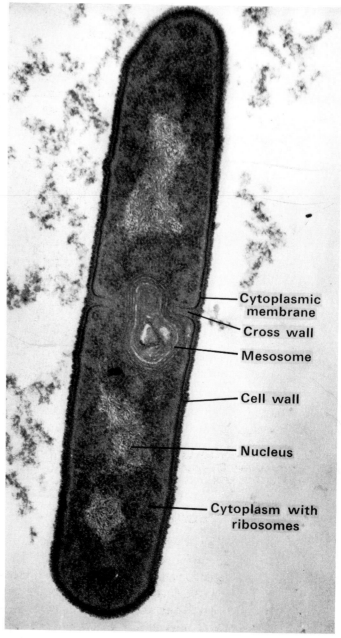

Fig. 2.2. A thin section of a dividing cell of *Bacillus licheniformis*. The mesosome is apparently involved in cross-wall formation. (Courtesy of P. Highton) (Mag. ×42 000).

(d)   A membrane, to maintain the internal environment.

(e)   Cytoplasm, as a solvent for the utilization of food.

Thus, all cells share a common chemical composition, the common chemical activities of metabolism and a common physical structure of organization, and together these give them the capabilities of growth and self-replication.

The prokaryotic cell

Prokaryotic cells can vary in size from a *Mycoplasma* (a sphere of about 0.12 μm diameter) to a blue-green bacterium like *Oscillatoria* (a rod of dimensions as much as 40×5 μm), but the majority have a diameter in the region of 1 μm. The components of a prokaryotic cell are shown in Figs 2.2, 2.3 and 2.4.

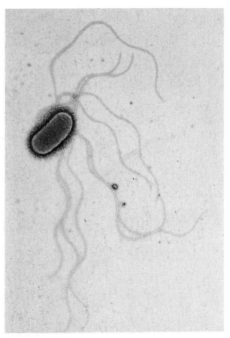

Fig. 2.3. Negative-stained preparation of a cell of *E. coli* with ten flagella. (Courtesy of D.W. Gregory) (Mag. ×10 000).

The cell membrane

The cell membrane is the boundary layer of the prokaryotic protoplast. In thin sections it can be seen as a triple-layered structure consisting of two electron-dense regions surrounding an electron-transparent one, a type of structure typical of most selectively permeable membranes in living organisms and called a 'unit membrane'. The chemical components of the cell membrane have been analysed using preparations made after controlled osmotic lysis of protoplasts. The main components are lipid and protein present in about equal

19

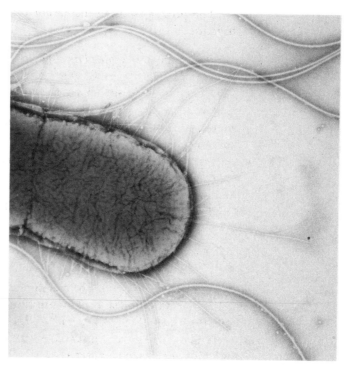

Fig. 2.4. Negative-stained preparation of a cell of *E. coli* with flagella and many pili. (Courtesy of D.W. Gregory) (Mag. ×50 000).

amounts in a 'fluid mosaic' structure (Fig. 2.5). This is composed of a bilayer sheet of phospholipid molecules, their polar heads on the surfaces and their fatty acyl chains forming the interior. The protein components are embedded within the phospholipids, some spanning the membrane, some on one side, some on the other. The membrane is fluid in that components can diffuse laterally in its structure. It is also asymmetric, and by its activities it creates and maintains gradients across itself, so that the inside of the cell is very different to the outside environment.

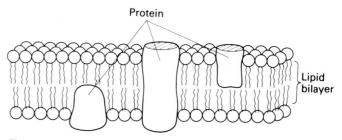

Fig. 2.5. Diagram of the 'fluid mosaic model' of a biological membrane, showing the bilayer of phospholipids and embedded protein.

20

The prokaryotic cell membrane can have the following functions:

1 *Selectively permeable layer.* The function of allowing the entry and exit of some molecules but not others is exceedingly important and if this barrier is broken down, essential metabolites pass out of the cell and the result is death. The transport of the molecules across the membrane in either direction usually involves their specific combination with protein molecules called permeases which are built into the membrane structure. Because of the specificity of this combination, a large number of different permeases may be required in any one cell. Energy provided by metabolism may be necessary for the transport process and considerably higher concentrations of solutes may occur within the cytoplasm compared with the external environment.

2 *Energy production.* The membrane is the site of electron flow in respiration and photosynthesis leading to the conversion of ADP to ATP, and is therefore the site of the enzymes and carriers involved in these reactions. Further details are given in Chapter 4.

3 *Extracellular polymer production.* The final stages in the synthesis of some of the polymers in the cell wall, capsule and extracellular fluids are catalysed by membrane enzymes. However, extracellular proteins including those in flagella and pili are formed on cytoplasmic ribosomes and are specifically transported across the membrane to the exterior of the cell.

4 *Site of chromosome attachment.* This important role is discussed in the section on the nuclear body.

In this discussion, it has been assumed that the cell membrane is a simple structure underlying the cell wall and following its contours. However, infoldings may occur which can produce complex internal structures and increase its surface area. These are seen for example in chemoautotrophic bacteria with high rates of aerobic respiration such as *Nitrosomonas*, and in photosynthetic bacteria, such as *Rhodopseudomonas*, where they are the site of photosynthetic pigments. Some electron micrographs of bacteria show localized infoldings of the membrane, the 'mesosome' (e.g. Fig. 2.2), which probably are the effect of chemical fixation on a specialized region of the cell membrane, in this case involved in septum formation, rather than being true structures *in vivo*.

21

Cytoplasm

The cytoplasm is sequestered within the cell membrane. It contains a variety of enzymes, coenzymes and metabolites and its main function is in intermediary metabolism and in providing an equable chemical environment for cellular activities. There is evidence that it has structure, i.e. it is not just like a solution of enzymes and metabolites in a test tube, but this concept is still poorly understood.

The nuclear body

It is possible to use coloured or fluorescent stains to delineate a central DNA-rich region in bacterial cells. This corresponds to the nuclear body which can be isolated from gently lysed protoplasts, when it is seen as a structure consisting of loops of supercoiled DNA together with about 10% RNA and 10% protein (mostly RNA polymerase). This nuclear body occupies about 10–20% of the cell volume, and in the electron microscope can show as a diffuse area containing fibrous material with no limiting membrane.

The DNA forms a single circular chromosome usually in the region of 1–2 mm (1 000–2 000 μm) in length. Since this is about a thousand times longer than the cell itself, the thread must be highly folded in the nucleus to give the bundles of fibres seen in thin sections. Unlike eukaryotes, discussed later, there are no histones in the nuclear material. Although a single chromosome has only been demonstrated in a few bacteria, genetic evidence of the presence of a single linkage group in other organisms suggests a probable universal occurrence in prokaryotes. Given a single chromosome, it is not necessary to have a complex mitotic system nor a nuclear membrane. Instead, there is an attachment between a specific point on the chromosome and the cytoplasmic membrane at which replication starts. The first stage in nuclear division involves duplication of this attachment site, followed by a progressive bidirectional replication of the DNA by two replication forks. The resulting two daughter chromosomes are then drawn apart by separation of these attachment points, and are induced to coil up to form the two new nuclear bodies (see Chapter 4). Sexual reproduction is rare in prokaryotes; when it does occur it is unidirectional and incomplete (see Chapter 6).

In contrast to the eukaryotic cell, the arrangement of the nuclear body in the prokaryotic allows transcription and translation to be closely coupled processes. Thus, translation of a messenger RNA molecule can occur while the mRNA is still being transcribed from the DNA. It is possible that

22

transcription and translation occur only at the interface between nuclear body and cytoplasm, so that genes have to be exposed there in order to be expressed.

Ribosomes

The ribosomes are seen in thin section as relatively dense particles about 20 nm in diameter. They can easily be prepared from ruptured cells by differential centrifugation on a sucrose gradient. They are made up to two sub-units of sedimentation constants 30S and 50S, which combine to give the characteristic prokaryotic 70S ribosome; both components are made up of roughly equal amounts of RNA and protein. The 30S subunit contains one RNA molecule of 16S, and the 50S contains two, of 5S and 23S. Some antibiotics such as streptomycin and chloramphenicol specifically inhibit protein synthesis by the prokaryotic ribosome.

In general, prokaryotic cells are characterized by a considerably higher rate of multiplication than eukaryotic cells and this difference is reflected in the number of ribosomes per unit mass. In rapidly growing bacteria, they may make up about 40% of the cell dry weight, chiefly as polysomes, which are assemblages of about 20 ribosomes actively translating the messenger RNA (mRNA) molecules.

Storage granules

Storage granules (sometimes called inclusion bodies) may occur within the cytoplasm of a cell. The main types of granules are glycogen, poly-β-hydroxybutyrate, polyphosphate and sulphur. Their number and size vary according to the cultural conditions, but in the presence of an excess of an external energy source they can make up as much as 50% of the dry weight. Under deficient conditions they are broken down to provide useful cell building blocks, energy or both.

The cell wall

The cell wall is the dense layer surrounding the cytoplasmic membrane. Before considering its structure and functions, it may be useful to digress and outline the most important differential staining method in microbiology which is used to separate bacteria into two fundamental groups—the Gram stain.

There has been much discussion concerning the biochemical basis of this staining method. It is probable that its differential staining relies on differences in the porosity of the

23

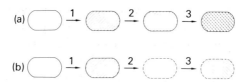

Fig. 2.6. The probable mechanism of the Gram stain. (1) Treat with crystal violet and iodine. A crystal-violet-iodine complex is deposited within the cells. (2) Treat with alcohol or acetone. In Gram-negative cells, the crystal-violet-iodine complex is leached from the cells because of the greater porosity of the cell wall. (3) Counterstain with basic fuchsin. Gram-positive cells (a) are doubly stained to give a purple colour while Gram-negative cells (b) stain with basic fuchsin only.

cells to the crystal–violet–iodine complex formed within them (Fig. 2.6). This difference can be related to wall structure. Viewed in thin sections, Gram-positive bacteria have a thick, amorphous single-layered structure (Fig. 2.2), whilst Gram-negative bacteria normally show a more complex multi-layered wall. In Gram-positive bacteria the main chemical component is the peptidoglycan (see below) and this thick layer prevents removal of the crystal–violet–iodine complex by alcohol or acetone. In the Gram-negative cell wall the peptidoglycan only makes up a small proportion of the structure, the remainder being protein, lipid, lipoproteins and lipopolysaccharides. The alcohol treatment is likely to increase porosity by removing lipid and this together with the thin layer of peptidoglycan gives rise to the differential staining. Let us consider the wall components in more detail:

1   *The peptidoglycan.* This is the most characteristic polymer in prokaryotes and is built up in a unique way by cross-linking polysaccharide chains by short peptides to give a vast macromolecule of the shape and size of the cell and of considerable mechanical strength. One of the most important aspects of the chemical structure is the presence of unique monomers in both the polysaccharide component (N-acetyl-muramic acid) and the polypeptide component (D-amino acids and sometimes diaminopimelic acid). Apart from providing resistance to degradative enzymes (except lysozyme and a few autolytic enzymes) this unique chemical structure is also responsible for providing the basis for the actions of many specific antibacterial agents, especially the β-lactams such as penicillin (see Chapters 8 and 10).

2   *The teichoic acid.* In Gram-positive cells, such as *Bacillus* and *Staphylococcus* species, the wall is about 50% peptido-glycan, and 50% teichoic acid, a unique polymer of a polyol phosphate, such as glycerol phosphate. The teichoic acids of different bacteria have a range of substituents, including amino acids, sugars and lipids, giving them individuality.

24

3   *The outer membrane.* In Gram-negative cells, such as *Escherichia* and *Salmonella* species, the thin layer of peptidoglycan is surrounded by the outer membrane. This contains some phospholipids making up a bilayer, but with much of the outer layer being of a unique polymer, lipopolysaccharide, of complex structure, which projects outwards. It also contains proteins, some associated with transport in and out of the cell, and lipoproteins, some of which extend inwards and anchor the outer membrane to the peptidoglycan.

4   *The function of the cell wall.* The main function of the wall is that of providing a mechanically strong bounding layer. Although a few prokaryotes do not have a cell wall, they can exist only in a very restricted range of protected habitats. An example of this is *Mycoplasma* inhabiting mammalian cells. The cell wall is not a semi-permeable membrane but it can act as a molecular sieve preventing large molecules passing through; in fact, particularly in Gram-negative cells, some enzymes and metabolites are trapped between the cell membrane and the outer membrane to form the *periplasm*. The wall components of the cell can be strongly antigenic (see Chapter 8). Thus, teichoic acids are the principal surface antigens of Gram-positive cells and lipopolysaccharides of Gram-negative cells.

The capsule

Some prokaryotes have a gel layer called the capsule surrounding the cell wall. It can be seen in the light microscope by negative staining with a particulate dye incapable of penetrating it such as Indian ink, but in the electron microscope it is normally only visible as an amorphous shrunken layer. The capsular gel is usually formed of a polysaccharide (1–2% dry weight) in water and there is a wide variety of different monosaccharide components joined in very many different ways. Occasionally, capsules are made of polypeptide gels such as the peculiar polymer of the unusual D-glutamic acid found in the bacteria which cause the disease of anthrax.

The function of the capsule seems to be mainly as a protective layer against attack by phagocytes and by viruses; it may also help to prevent too rapid and lethal a loss or gain of water in the recurrent dehydration and hydration that occurs in many habitats. Finally, the capsule usually has an ion-exchange capacity which may aid in the concentration and uptake of essential cations.

Flagella and locomotion

Most motile bacteria possess long, thin (c. 20 nm diameter) helical extracellular appendages called flagella (Figs 2.3 and 2.4) which are attached at one end through the cell wall to the cell membrane by a special terminal hook and basal body. The individual flagellum is not visible using the light microscope without increasing its effective diameter by coating it with a suitable precipitate. In the electron microscope, negative-staining with phosphotungstic acid shows the flagellum to be made up of identical sub-units arranged helically along the axis of the flagellum to give a hollow tube. These sub-units can be separated from each other by acidification and consist of protein molecules called flagellin; neutralization can cause automatic reaggregation to give a flagellum-like structure, a process presumably analogous to that occurring normally during flagellar growth. The arrangement and number of flagella on a cell can be a useful criterion for identification and classification.

The flagellum rotates, being driven by a rotary motor in its basal body. The hook acts as a universal joint, and the flagellum acts similarly to a ship's propeller, projecting backwards and driving the cell forward by exerting a viscous force against the aqueous medium. The cell occasionally changes direction by the motor briefly going into reverse, when the cell instantly stops swimming to restart again in a random direction when the motor switches back to its driving sense. Cells with many flagella do not get them entangled; in the driving sense, each one rotates individually, but they come together to form a stable bundle. On reversal they disperse and the cell tumbles until their motors synchronously switch back and they again form the stable bundle, and the cell swims off again.

The function of flagella is in locomotion and all naturally occurring flagellate bacteria are motile. However, there are other less common types of motility in prokaryotes.

1 *Gliding movement*, for example by some blue-green bacteria and myxobacteria, which requires contact of the cell with a surface, and probably involves some sort of contractile element built into the outer layers of the cell.

2 *Spirochete movement*. The spirochetes, a special group of bacteria including *Treponema*, the causative agent of syphilis, have helically shaped cells with axial filaments like flagella extending back from either end of the cell, but enclosed within an outer sheath. The rotation of the filaments 'periplasmic flagella', between the cell and sheath, results in a

26

helical wave propagating through the spirochete which thus swims through the medium.

The function of motility in a bacterium is so that it can respond to environmental stimuli, for example by swimming towards nutrients. It does this by a 'biased random walk'. It senses the stimuli at regular intervals, and if they increase in intensity, the cell swims for a longer time in a straight line before the flagellar motors reverse and it tumbles. If they decrease, it swims for a shorter time. By this trial and error method, it swims up a concentration gradient of an attractant. Similar but opposite responses result in its swimming down a concentration gradient away from an unfavourable environment. In the case of chemicals, these responses are mediated by specific chemoreceptors at the cell surface.

Pili

Prokaryotic cells may have appendages called pili which look superficially similar to flagella (Fig. 2.4). They too are built up of individual protein sub-units of pilin, arranged helically to form a filament. However, they differ from flagella in several important respects:

1   The filament is usually straight and is shorter than a flagellum.
2   The diameter is smaller (about 10 nm).
3   The function is not in motility. Some, the f-pili, occur on male cells and act as a bridge between conjugating cells (Chapter 7). Others serve for attachment of cells to substrates, e.g. pathogens to their hosts, for example *Neisseria gonorrhea* to cells of the human urinary tract. Both flagella and pili are antigenic and can be specific sites of attachment for bacteriophages.

Spores

Some prokaryotes, especially *Bacillus* and *Clostridium* species inhabiting soil, produce structures called spores. These have immense resistance to environmental stresses such as heat, desiccation and radiation. Thus, the spore serves to prolong the life of the cell in environmental conditions which are no longer conducive to growth.

Characteristically, a single endospore is formed within a vegetative cell and on germination a single vegetative cell is again produced.

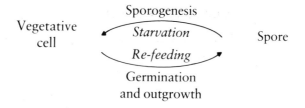

Vegetative
cell

Sporogenesis
Starvation
Re-feeding
Germination
and outgrowth

Spore

There has been much interest in this process since it provides a very simple model for the differentiation of one type of cell into another and could provide clues to the more complex processes occurring in animals and plants.

In the phase-contrast microscope, spores appear as highly refractile bodies sometimes greater in diameter than the cell from which they were formed. Thin sections in the electron microscope show a complex multi-layered wall (Fig. 2.7).

Inside an exosporium of variable structure, there is a spore coat composed of several laminated layers of protein. Below this is the thick cortex containing a specific peptidoglycan and, below that, the protoplast containing the most characteristic chemical component of the spore—a complex of calcium and dipicolinic acid which is thought to contribute to heat resistance.

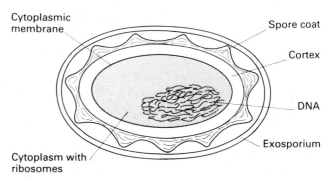

Fig. 2.7. The structure of a typical bacterial endospore.

Although the endospore is the typical resting stage in prokaryotes, two other forms can occur:

1 *Cysts.* These are intermediate in structure and resistance properties between an endospore and a vegetative cell. An example of cysts in myxobacteria is seen on page 43.

2 *Conidia (exospores).* These are produced by fragmentation of the aerial ends of the multinucleate hyphae of the Actinomycetes. Because of the similarity to the mycelial structure and conidia formation in fungi, the Actinomycetes used to be classed as fungi although it is now clear that they are prokaryotic cells in a multi-nucleate and branched hyphal form.

The eukaryotic cell

Eukaryotic cells are generally much more complex in structure than prokaryotic cells. They are normally much larger, with a typical diameter ten times greater than that of the prokaryote (i.e. 10 μm) and they show a very great diversity in size and shape. It is difficult to define a typical eukaryotic cell structure and it would probably be misleading to do so;

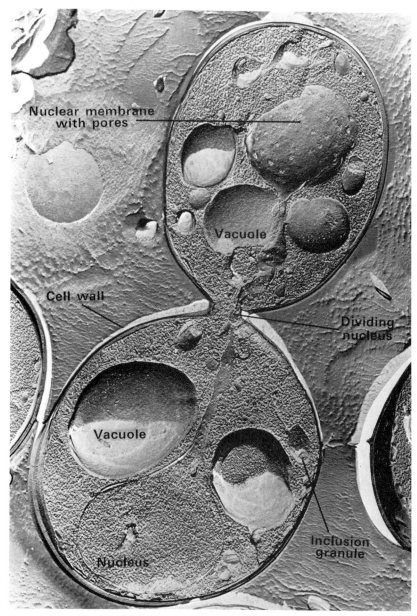

Fig. 2.8. A freeze etched preparation of the yeast *Saccharomyces cerevisiae*. The nucleus is seen dividing with a connection between the separated nuclei. (Courtesy of H. Moor) (Mag. ×15 000).

the yeast presented in Fig. 2.8 should only be taken as an example of one type. Since most students will have an idea of the structure of some eukaryotic cells from their studies of botany and zoology, only some of the components will be mentioned in any detail and an emphasis will be placed on those structures that serve to distinguish the cells fundamentally from those of prokaryotes.

Membranous structures

A variety of membranous selectively permeable structures occur within a eukaryotic cell, giving a complex multi-compartmental whole.

1 *The plasma membrane.* In both physical and chemical structure, the plasma membrane is very similar to the cytoplasmic membrane of prokaryotes. However, a difference is the presence of sterols in eukaryotic membranes. The eukaryotic plasma membrane has the same vital function as the prokaryotic cell membrane in being selectively permeable and of transporting specific solutes, but it does not have any properties of respiration or photosynthesis. An additional property however, not possessed by prokaryotes, can be the ability by wall-less cells to ingest food in particulate form by phagocytosis or in liquid form by pinocytosis; in either case, a membrane-enclosed vacuole is formed within the cytoplasm (Fig. 2.9).

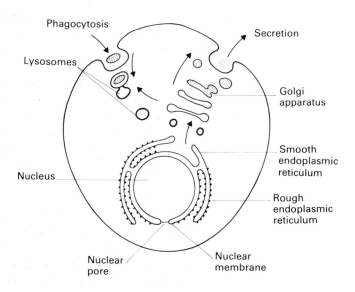

Fig. 2.9. Interrelationships between membranous structures in a eukaryotic cell.

2 *Endoplasmic reticulum.* The endoplasmic reticulum typically comprises more than half of the membrane in the cell. Topologically it is a sac, with the cytoplasm outside and its lumen inside. Its ramifications through the cell give it a large surface area. In electron micrographs two forms are distinguishable:

(a) *Rough ·endoplasmic reticulum* is studded with ribosomes on its outer surface. This is the site of the synthesis of proteins destined for export from the cell, or to become wall components or to be stored in vesicles or vacuoles within the cell, such as the lysosomes which contain digestive enzymes. As they are synthesized these proteins are secreted into the lumen of the endoplasmic reticulum, often becoming glycosylated to glycoproteins by membrane-bound enzymes.

(b) *Smooth endoplasmic reticulum* lacks ribosomes. It is the site of phospholipid and sterol synthesis in the membrane, and the site of formation of membrane vesicles which are transferred to the Golgi apparatus for further processing.

3 *Golgi apparatus.* This is a series of membranous sacs, originating from the endoplasmic reticulum. It is a major site of biosynthesis of polysaccharides that are destined for export from the cell, such as pectin-like wall components in algae. Proteins and glycoproteins formed in the endoplasmic reticulum are covalently processed, for example by further glycosylation, while in the Golgi apparatus. The outer sacs of the Golgi body then release these products in vesicles, some of which move to the plasma membrane, fusing with it, releasing their contents and adding to its surface area. Other vesicles are stored in the cell, for example digestive lysosomes.

4 *Nuclear membrane.* The nucleus is surrounded by membrane which is in continuity with the endoplasmic reticulum, and is thus topologically a membranous sac enclosing a lumen. The inner-facing membrane has specific proteins on its surface which are thought to hold the shape of the nucleus and to bind specific regions of chromatin. The nuclear membrane has pores, allowing transport of molecules in and out of the nucleus. Thus, messenger RNA must be transported out and enzymes such as RNA polymerase must be transported into the nucleus.

5 *Vacuoles, lysosomes and peroxisomes.* As mentioned above, phagotrophic micro-organisms such as many protozoa engulf food particles by invagination of the plasma membrane. Digestion takes place by the fusion of a lysosome with this food vacuole. Vacuoles are formed in many cells, for example yeasts, and appear to be involved in the accumulation and storage of metabolic intermediates and ions. The contractile vacuoles occurring in some cells, e.g. freshwater

protozoa, function in osmotic regulation and in the excretion of waste products. Peroxisomes are membrane-bound packages of oxidative enzymes such as catalase. They are formed by budding off from smooth endoplasmic reticulum.

Thus, these membrane systems must be thought of in terms of their spatial and temporal relationships; they are common parts of a dynamic system of membrane flow and modification (Fig. 2.9).

Mitochondria and chloroplasts

These two organelles have many features in common:

(a) They are both concerned with energy conversions. Mitochondria are the site of oxidative phosphorylation, converting the energy of oxidative reactions into useful forms, especially ATP. Chloroplasts are the site of photosynthesis, utilizing light energy to produce ATP and NADPH, and using these to fix carbon dioxide to carbohydrate.

(b) They are both surrounded by two membranes. The outer one is relatively permeable and resembles the cell membrane, for example by having sterols. The inner one is impermeable and resembles the bacterial cell membrane.

(c) Their internal bioenergetic membranes have a large surface area. In the mitochondrion, this is the inner membrane which is extensively infolded to form cristae on whose surface are the respiratory chain enzyme complexes. In the chloroplasts, within the inner membrane are stacks of membranous sacks (thylakoids) bearing the photosynthetic pigments and associated acceptors.

(d) The matrix within the inner membrane in each case contains many enzymes; for example in mitochondria those for the citric acid cycle, and in chloroplasts the ribulose bisphosphate carboxylase, the enzyme responsible for fixing carbon dioxide which can be so concentrated in the chloroplasts of many algae that it forms the most obvious object in the cell, the pyrenoid.

(e) They both have a quasi-independence in the cell as they both have their own DNA and transcription and translation machinery. These resemble those of bacteria, as the DNA is several copies of a single circular chromosome, and the ribosomes are similar in size, make-up and sensitivity to antibiotics such as streptomycin and chloramphenicol. Their protein synthetic ability, however, is limited to only a few of their essential components, for example sub-units of cytochrome oxidase and ribulose bisphosphate carboxylase. Other sub-units and essential components are coded for by the cell nucleus and translated on cytoplasmic ribosomes to

32

be transported into the mitochondrion or chloroplast.

(f) Following from (e), mitochondria and chloroplasts have to replicate their DNA and divide in pace with their parent cells. This can be prevented by specific antibiotics, so that treatment of the alga *Euglena* with streptomycin prevents its chloroplasts dividing, changing it from a green photoautotroph to a colourless chemoheterotroph.

(g) Following from (e) and (f) has come the idea that eukaryotic cells arose by endosymbiosis, the putative fermentative pro-eukaryote engulfing an aerobic bacterium and then a blue-green bacterium. This hypothesis is of course unprovable, but circumstantial evidence comes from the observation that endosymbiosis is a common phenomenon; for example, some strains of *Paramecium* contain symbiotic bacteria and others contain symbiotic algae which divide in pace with their hosts.

The cytoplasm

The eukaryotic cytoplasm, like that of bacteria, contains many enzymes, metabolites and other solutes, but is also structured as it contains a cytoskeleton composed of microtubules, actin, myosin and other structural proteins. Cytoplasmic streaming is common, as seen by the mass movement of organelles within cells.

Eukaryotic cytoplasmic ribosomes, although having the same function of translation of mRNA to protein as in prokaryotes, differ in having a larger sedimentation constant (80S) made up of sub-units of 60S and 40S, and a different sensitivity to antibiotics, being inhibited by cycloheximide but not by streptomycin and chloramphenicol.

The nucleus

In eukaryotic cells, the nucleus is a definite structural entity surrounded by a membrane and containing many chromosomes. The individual chromosomes are made up of linear molecules of DNA compared with the circular DNA of prokaryotes. Associated with the nucleus there is normally an RNA-containing body called the nucleolus which is a specialized structure responsible for ribosomal-RNA synthesis, a function carried out by the single chromosome of the prokaryotic nucleus. The eukaryotic chromosomes also differ in having their DNA complexed with basic proteins called histones to form chromatin. As the chromosomes contain much more DNA than does the prokaryotic chromosome, each is replicated by many replication forks, acting in pairs from many sites of origin. This DNA doubling is followed by the complicated process of mitosis designed to ensure an orderly partition of a complete set of chromosomes to each daughter cell.

Sexual reproduction is common in eukaryotic micro-organisms, and the consequent doubling of the haploid chromosome number then requires meiosis. This allows organisms to have alternate haploid and diploid life cycles (see Fig. 7.11).

Storage granules

Eukaryotic microbial cells contain a variety of inclusion bodies, such as membrane-bound granules of starch, protein or lipid droplets.

The cell wall

The cell walls of eukaryotic micro-organisms vary widely in shape, thickness and chemical composition. Indeed, some protozoa apparently do not have walls at all, although it is probable that they must have some additional strengthening of the plasma membrane to maintain cell shape and rigidity. When walls do occur, they can be isolated by a similar procedure to that mentioned previously (p. 17) and usually prove to have a simpler structure than those of prokaryotes.

1   *Algae.* The basic structure is usually maintained by microfibrils formed by the intertwining of long polysaccharide molecules, such as cellulose, mannans, xylans and pectins, while there are other algae with silica or calcium carbonate walls often sculptured into fascinating and beautiful shapes.

2   *Fungi.* The common structural wall polymer is chitin ($\beta(1-4)$poly-N-acetylglucosamine), often occurring with polyglucose built of $\beta(1-3)$ linkages compared with $\beta(1-4)$ linkages of cellulose. The mechanical strength of the wall is maintained by cross linkages between microfibrils of chitin and the glucan matrix to give a thick and tough structure. Mannans are also common in yeast walls.

3   *Protozoa.* There is a great variety of surface components in protozoa, including structures built from protein, cellulose, calcium carbonate or silica.

Flagella, cilia and locomotion

The movement of eukaryotic micro-organisms is usually by the action of flagella and cilia which have a structure quite different from that of the prokaryote flagellum. They are composed of a characteristic '9 plus 2' arrangement of microtubules surrounded by a sheath which is an extension of the plasma membrane (Fig. 2.10). They are powered by ATP. Another method of locomotion in eukaryotic micro-organisms is amoeboid movement, a result of cytoplasmic streaming in cells without a cell wall. This is the result of interactions between actin and myosin in the cytoskeleton. Similarly to prokaryotes, motile eukaryotes show movement towards or away from heat, light or certain chemical substances.

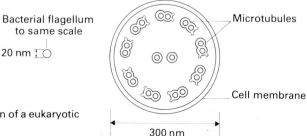

Fig. 2.10. Cross-section of a eukaryotic flagellum.

Spores

Eukaryotic micro-organisms show a great variety of spores, some produced in vast numbers for dispersal, others produced in small numbers for survival. These latter tend to be thick walled and resistant to environmental stresses.

Comparison of prokaryotic and eukaryotic cells

The result of the structural studies outlined here has been to show that prokaryotic and eukaryotic cells differ in many aspects. These are summarized in Table 2.1.

Table 2.1. A comparison of the main distinguishing features between prokaryotic and eukaryotic cells

	Prokaryotic cell	Eukaryotic cell
Typical diameter	1 μm	10 μm
Nuclear membrane	no	yes
Chromosome	one	more than one
DNA	circular	linear
Histones	no	yes
Mitosis, meiosis	no	yes
Endo− and exocytosis	no	yes
Sexual reproduction	rare, incomplete	common, complete
Site of oxidative phosphorylation	cell membrane	mitochondrion
Site of photosynthesis	cell membrane	chloroplast
Cytoplasmic ribosome size	70S	80S
Peptidoglycan	usually present	absent
Flagellum	1 flagellin fibril	'9+2' microtubules and membrane
Amoeboid movement	no	yes

3 : A Survey of Micro-organisms

Before considering the vast range of micro-organisms occurring in nature, a general word must be said about the patterns of nomenclature, identification and classification. The object of the nomenclature is simply to be able to give an organism a name, whilst the object of classification or taxonomy is to arrange the named organisms into orderly groups that reflect the similarities within a group and the differences from other groups. If we are solely concerned with identification as, for example, in the diagnosis of a disease, then the characters employed in this identification should be easy to measure, should be reproducible in a variety of laboratories and should provide identification with as few characters as possible. In other words, we want to produce a key which is valuable with respect to the original purpose of a practical identification. Most microbial taxonomy is of this type and inevitably some characters are considered more important than others simply because they are more useful.

Numerical taxonomy uses as many characters as possible, each of which is of an equal weight. The results are fed into a computer so as to define the similarities and differences between micro-organisms and thus to indicate possible natural groupings. As microbial fossils can provide little detailed evidence, a **phylogenetic taxonomy** based on the evolutionary relationship of micro-organisms is only now becoming possible with techniques of molecular biology.

Let us consider some of the characters that are used by the microbiologist:

1 *Morphological characters.* These concern cell shape and size, staining reactions, presence or absence of spores or reproductive forms, type of motility, etc.

2 *Cultural characters.* These include the cultural requirements for multiplication (e.g. nutrients, oxygen, temperature,
etc.) and the way growth occurs in liquid media, and particularly on solid media (e.g. colony form).

3 *Biochemical characters.* More specific biochemical properties such as the metabolic end-products and the presence
or absence of a particular enzyme or pathway.

4 *Serological characters.* The nature of the surface antigens
as revealed by suitable specific antibodies.

5 *Molecular characters.* Ultimately, the evolutionary
history of an organism is recorded in the sequences of bases in
the DNA. Our increasing knowledge of molecular genetics is
allowing us to unravel some of this history. (*a*) *GC ratios*: the
variable part of DNA is its sequence of bases. The four kinds
of bases are adenine (A), hydrogen-bonded in double-
stranded DNA to thymine (T), and guanine (G) to cytosine
(C). Thus, the amount of A equals that of T and the amount of
G equals that of C. However, there is a considerable amount
of variation in the ratio of $G+C$ to $A+T$. Gross comparisons
of 'GC ratios' are thus readily made and can indicate gross
relationships between bacteria. (*b*) *DNA–hybridization*: the
extent of hybridization between single-stranded DNA of two
different micro-organisms provides a measure of very close
relationships that may exist between them. (*c*) *RNA
sequences*: the sequencing of bases in appropriate sections of
DNA provides an excellent means of comparing bacterial
strains, but still presents great technical difficulties. An
alternative is to sequence the smaller molecules (RNA or
protein) derived from DNA, one of the most useful techniques for which is proving to be the comparative sequencing
of specific RNA molecules. These molecules are highly con-
served in an evolutionary sense and provide a very valuable
measure of the phylogenetic relationships of the wider
groupings of micro-organisms. The chosen molecules are
ribosomal RNA components, the 16S rRNA and the 5S
rRNA.

Granted some sort of taxonomic system, how should a
micro-organism be named? It is usual to use a binominal
system; each distinct species is given a name consisting of two
words—the first is the genus and is written with a capital
letter while the second is the specific epithet (the species) and
is not capitalized. Sometimes species are further divided into
varieties or strains, e.g. *Escherichia* (genus) *coli* (species) K12
(strain). It is evident that once even a simple binominal
nomenclature is used, an hierarchical scheme is implied and
three questions must be asked:

1 How is a species defined?

2   How many species should be grouped into a single genus and what should the criteria be for this grouping?

3   Should genera be similarly grouped to produce a multi-tiered hierarchical classification?

Let us consider these questions briefly with particular respect to bacteria.

(a)   *The species.* In higher organisms the inability of two organisms to undergo sexual reproduction is used to define separate species. As is obvious from its nature (Chapter 7), sexual reproduction cannot be used as a criterion for bacterial speciation. Further, micro-organisms are commonly haploid and are often subject to high selective pressures so that the same species isolated from different locations may exhibit subtle differences (Chapter 7). Thus, it is necessary to take a useful selection of common characters and use them to define a species.

(b)   *The genus.* Likewise, a grouping of species into a genus must be a practical one in the absence of phylogenetic information. Again, a grouping of similar characters must be used, more restricted and possibly more fundamental than that of the species. It must be admitted that the very vagueness of the concept of genus has led to vast differences in the way it is used. Thus, the genus *Bacillus* includes all aerobic, rod-shaped, endospore-forming, Gram-positive bacteria and a wide range of species are represented; on the other hand, the genus *Salmonella* is restricted to a group of bacteria pathogenic for man and other animals, and apart from the type of disease caused and the specific chemistry of surface antigens, these differ very little. Rather to the surprise of the more cynical microbiologist, the use of the computer in numerical taxonomy has given support to the concept of species and, to a lesser extent, of genus. In other words, bacteria do fall into natural groupings although the reason for this may well be that they inhabit similar environmental niches to which they have become adapted. This adaptation has in itself selected certain characteristics.

(c)   *Families*, etc. It is common to group genera together and so on up through an hierarchical taxonomy with the following divisions: Species, Genera: (tribes); Families: (suborders); Orders; Classes. It is, of course, attractive to the tidy-minded to produce such a system, but since it cannot aid the identification or nomenclature of an organism, the only real justification must be a phylogenetic one. We are now reaching the stage where this should be possible to accomplish, both with bacteria and with the eukaryotic micro-organisms.

Living organisms are best thought of today as being members of one of the five Kingdoms:

38

1   *Monera* (Prokaryotes, bacteria), consisting of the Archaebacteria (see below), the Oxyphotobacteria, i.e. oxygen-
yielding photosynthetic blue-green bacteria (the cyanobacteria), and the Eubacteria. More than 5000 species
described.

2   *Protista*, consisting of algae, protozoa, slime moulds and
water moulds. More than 12000 species described. The slime
moulds and water moulds have traditionally been studied by
mycologists, and so are usually considered as fungi, but
phylogenetically they are clearly protists.

3   *Fungi*, consisting of the zygomycetes, ascomycetes,
basidiomycetes and deuteromycetes. More than 60000
species described.

4   *Plantae*, consisting of bryophytes and tracheophytes.
More than 250000 species described.

5   *Animalia*, consisting of invertebrates and vertebrates.
More than one and a half million species described.

Of these, the first three Kingdoms (together with the viruses)
form the province of the microbiologist. Although not so
numerous in species, the micro-organisms of the biosphere
vastly outnumber and also probably outweigh the plants and
animals in the biosphere.

Prokaryotic micro-organisms

How can we subdivide the prokaryotes? Although this is
often done, we have already pointed out the artificiality of
formally doing so given our present knowledge. Some groupings will be mentioned (e.g. myxobacteria, nitrogen-fixing
bacteria, spirochaetes) without any phylogenetic implications on the reality of such a grouping beyond the important
one of convenience.

Let us consider the range of prokaryotic micro-organisms
in terms of their structure, leaving the more biochemical and
genetic aspects to the appropriate chapters.

Cell shapes

There are three basic cell shapes: the sphere (coccus); the
rod (bacillus), a cylinder with rounded ends; and the curved
rod, either as one slight curve (vibrio) or as a spiral or helix
(Fig. 3.1).

Multicellular structures

Sometimes cell division and cross-wall formation are not
followed by the separation of the daughter cells. In this way

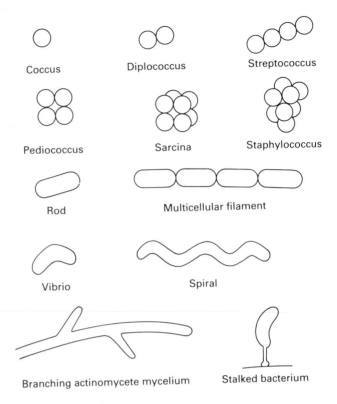

Fig. 3.1. Range of structures of bacteria.

an undifferentiated multi-cellular structure is produced, the shape of which will depend upon the planes of cell division. Thus, rods always divide in one plane, and, if the cells remain attached, a multicellular filament is formed.

In spherical cells a variety of shapes can occur:

1 cells divide in one plane and remain predominantly attached in pairs, e.g. *Diplococcus*,

2 cells divide in one plane and remain attached to form chains, e.g. *Streptococcus*,

3 cells divide in two planes to give plates, e.g. *Pediococcus*,

4 cells divide in three planes regularly to produce a cube, e.g. *Sarcina*,

5 cells divide in three planes irregularly producing bunches of cocci, e.g. *Staphylococcus*.

Spiral bacteria are predominantly unattached but the individual cells of different species show striking differences in length and in tightness of spiral.

In all these instances, the multicellular form is made up of separate individual cells. However, in actinomycetes multi-nucleate cells without cross walls produce branched mycelia of indefinite length which appear superficially similar to those of filamentous fungi.

Spores and cysts

Spores and cysts have already been discussed. Their occurrence and their position in the cell are often used as distinguishing characters.

Motility

As discussed previously, prokaryotic motility may be by flagella, by gliding or by axial filaments. If it is by flagella, their arrangement may be a useful diagnostic feature, i.e. are they at one end (polar), or distributed around the cell (peritrichous)?

Stalks

A few bacteria have a stalk by which they attach to a solid substratum using a holdfast at the tip. This attachment is usually a stable one and on cell division a flagellate cell is produced which swims around, eventually settling on a new surface where it forms a stalk in place of the flagellum. Thus, it has a primitive life cycle ensuring an attached and a swarming phase.

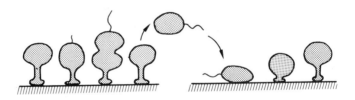

Succession of development

Buds

The majority of bacteria divide by symmetric division in which the cell gradually doubles in size and divides into two equal halves, each of which goes on to repeat the process. However, a few divide by budding, a process analogous to that characteristic of yeasts, in which growth is polarized and cell division is asymmetric (see Chapter 5). An interesting feature of symmetric division as distinct from budding is that there is no parent and offspring in the conventional sense and ageing apparently cannot occur.

41

Absence of cell wall

Some bacteria, usually called mycoplasmas, are incapable of forming a peptidoglycan cell wall. They are consequently highly pleomorphic because of the lack of wall rigidity and are prone to lysis even though the cytoplasmic membrane may be partially strengthened by substances like sterols derived from the host. Mycoplasmas are probably much more widespread in nature than was thought and they can commonly be isolated from warm-blooded animals including man as well as from plants. Since they have probably evolved with a loss in wall-forming ability in a variety of bacteria, they differ in many important characters. This illustrates the danger of grouping organisms on the basis of a single character.

Intracellular parasites of eukaryotes

A variety of bacteria can parasitize eukaryotes. The most interesting are those which have lost the capacity for growth outside the host cell. They show a primitive biosynthetic machinery and it is assumed that their evolution into a strictly parasitic existence has allowed a considerable simplification in the enzymes, and therefore genes, required. As a result of this simplification the cells are much smaller than independent bacteria, and may have a diameter of only about 0.3 μm. This may well be about as small as a cellular form of life can be since there is a limit to the degree of reliance that can be placed on the host cell without becoming a viral organization. Certainly organisms of the type considered here (rickettsias and chlamydia) contain both DNA and RNA and have most of the characteristics of prokaryotic cells.

Intracellular parasites of prokaryotes

There is a group of small, highly motile bacteria that are parasites of other prokaryotes. They adhere to the wall of the host, penetrate through to the periplasm where they replicate and eventually cause the lysis of the host cell. Organisms like this are called bdellovibrios and are probably common in environments such as soil.

Fruiting bacteria

A few bacteria called myxobacteria have the property of forming fruiting bodies, specialized multicellular structures reminiscent of some of those produced by slime moulds.

Under appropriate conditions such as nutrient deficiencies, a swarm of vegetative cells aggregate and form a fruiting body with a shape characteristic of the species. Some of the cells at the tips of the fruiting body then undergo differentiation to produce cysts.

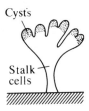

These fruiting myxobacteria present the most complex behavioural patterns and life cycles found in prokaryotes. Much interest has been shown in them since the process of fruiting body formation could provide a primitive and simple system for studying cell differentiation.

Blue-green bacteria

These organisms, frequently referred to as blue-green algae, have the typical cell structure of a prokaryote, as described earlier, but form a discrete grouping of bacteria which evolve oxygen during photosynthesis and have chlorophyll a, a pigment also found in all algae. They are very ancient, being recognizable in fossils dated 3.5×10^9 years ago. To put this figure in perspective, the age of the earth is 4.5×10^9 years, and the first recognizable eukaryotic cells are from 1.5×10^9 years ago. Blue-green bacteria are probably major primary producers in the world's oceans, and many are ecologically important as fixers of atmospheric nitrogen (see Fig. 9.1).

Actinomycetes

The actinomycetes are a large group of filamentous bacteria which characteristically show branching patterns just like those of fungi, to give rise to a spreading mycelium. Also like fungi, they often readily produce spores when grown in culture. They are easily isolated from soil, where they may play a role in the maintenance of crumb structure. Most soil isolates, in particular the genus *Streptomyces*, produce antibiotics, and so are of great importance to the pharmaceutical industry (see Chapter 8).

Archaebacteria

An exciting recent discovery is that an apparently disparate assembly of prokaryotes, previously classified in a range of bacterial groups, have the following features in common whereby they differ from other bacteria:

1   Their 16S rRNA molecules are similar to each other, but differ greatly from those of other bacteria and from eukaryotes.

2   Their walls do not contain peptidoglycan, but a range of other unique polysaccharides.

3   Their cell membrane is of a single layer of glycerol-hydrocarbon—glycerol chains instead of a bi-layer of phospholipids arranged tail to tail.

4   Their ribosomes are insensitive to chloramphenicol.

These prokaryotes also have another common feature in that they are said to inhabit 'extreme environments'. They include the methanogens, the methane-generating bacteria of anaerobic muds; the extreme red halophiles, the salt-loving bacteria of saturated brine and salted fish; and the thermo-acidophiles, found in hot sulphur springs or smouldering coal wastes. It is suggested that they represent a very ancient lineage which diverged from the eubacteria very early in the evolutionary process, and have survived only in these specialized ecological niches.

Eukaryotic micro-organisms

The eukaryotic micro-organisms show an immense range of structure, function, behaviour and habitat preference. As such they are clearly well worth studying in their own right, but some of them can be singled out and studied in particular detail as models for similar or identical activities occurring in 'higher organisms'. Pre-eminent among these in recent years has been the bakers'/brewers' yeast *Saccharomyces cerevisiae*, which has been a major model for eukaryotic molecular biology.

The protista

1   *The algae*

It would be difficult to underestimate the importance of algae to freshwater or marine environments. They can occur in such large numbers in the surface layers that the water appears coloured. In the oceans where they form a major part of the plankton, the total mass of algae is probably greater than that of all our landplants, and they are responsible for over half of the overall rate of biosynthesis of organic compounds from CO_2. Although algae mainly occur in waters, some grow in soil or on the surface of vegetation provided that the environment remains reasonably moist. In drier situations, algae have to be protected from desiccation, for example by association with fungal hyphae in the formation of lichens.

What are the characteristic properties of algae?

(a) They obtain their energy by an oxygen-producing photosynthesis occurring in chloroplasts. Within the chloroplasts, which show a great diversity in shape and number per cell, are found a variety of chlorophylls and carotenoids

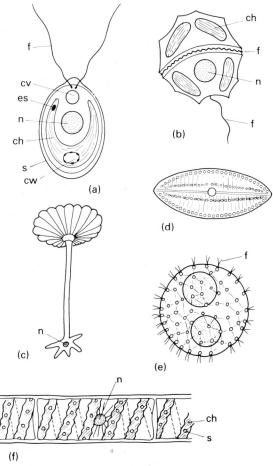

Fig. 3.2. Some algae. a–d: Unicellular forms; (a) *Chlamydomonas*; (b) *Peridinium*; (c) *Acetabularia*; (d) a diatom. (e) A colonial alga, *Volvox*, containing two daughter colonies; (f) a filamentous alga, *Spirogyra*. ch, chloroplast; cv, contractile vacuole; cw, cell wall; es, eye spot; f, flagellum; n, nucleus; s, starch grain.

which are often characteristic of the algal group. Chlorophyll *a* is always present.

(b)   They exhibit a wide range of morphological types (Fig. 3.2). Although some are unicellular, others produce cell aggregates embedded in mucilage; still others produce filaments which may be multicellular or coenocytic. Some algae such as the seaweeds have a very complex, almost plant-like colonial structure, although there is no cellular differentiation in the sense of its occurrence in higher animals and plants. It should be noted that some brown algae like the oceanic kelp are far from being microscopic and may reach a length of 50 m.

Some of the single-celled algae are fascinating in the size and complexity of a single cell. Thus, an *Acetabularia* cell has a root-like base, a stalk several centimetres long and an umbrella-shaped cap. It has been used as a test organism in studying nuclear-cytoplasmic interactions, as the very large nucleus is located in the base and can be easily removed. Nucleated portions can regenerate a complete structure but even a non-nucleated region can regenerate a stalk and cap provided that a reasonable amount of cytoplasm remains; this shows that stable cytoplasmic components (presumably long-lived mRNA) can control regeneration.

(c)   Many algae are motile, usually by flagella. In a colonial alga such as *Volvox* as many as 50 000 individual flagellated cells may make up a complex structure in which the action of the flagella is highly co-ordinated throughout the colony. On the other hand, diatoms glide along on a trail of mucilage that they exude.

(d)   Algae characteristically have a thick cell wall of poly-saccharide made of components such as pectin, cellulose or xylan, which is sometimes calcified with calcium carbonate. An important wall component for microbiologists is the agar gel of some red seaweeds. Some of the most fascinating walls are the silica shells of diatoms which exhibit a wide variety of beautiful shapes. These silica shells remain after the death of the organisms and they can accumulate as vast fossil deposits of so-called diatomaceous earth which may be as much as 1 000 m thick.

(e)   Algae can reproduce either sexually or asexually and can show very complicated life cycles. About 30 000 species of algae have been described.

2   *The protozoa* (Fig. 3.3)

The protozoa are commonly defined simply as unicellular animals. This is not a very helpful definition for the microbiologist and perhaps it would be better to say that they are a group of unicellular, non-photosynthetic eukaryotic micro-organisms which normally obtain their food by phagocytosis and which possess no true cell wall. A few of their characteristics will now be considered.

(a) The type of *movement* can be used to divide the protozoa into major groupings:

    1   Amoeboid motion is characteristic of organisms similar to *Amoeba*. Cytoplasm flows forward into a pseudopodium which is produced in the direction of movement on a solid surface; the opposite end of the cell is correspondingly retracted.

    2   Flagellar movement occurs in the flagellate protozoa, some of which are the colourless counterparts of

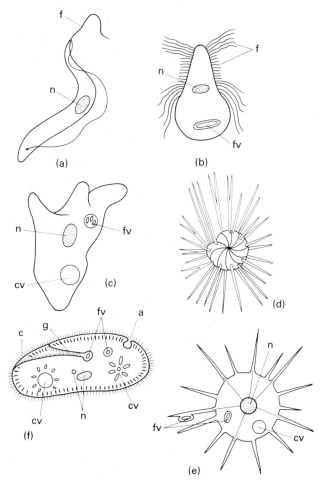

Fig. 3.3. Some protozoa. Flagellates; (a) trypanosome;
(b) *Trichonympha*. Amoeboid forms: (c) *Amoeba*; (d) a foraminiferan;
(e) a heliozoan. A ciliate: (f) *Paramecium*. a, anal pore; c, cilia;
cv, contractile vacuole; f, flagellum; fv, food vacuole; g, gullet;
n. nucleus.

47

particular algae. Since chloroplasts have their own DNA, their loss by a cell will be irreversible and a non-photosynthetic form which would be normally classed as flagellate protozoa will result. This process illustrates the hazy borderlines between the major groups of eukaryotic micro-organisms. Some protozoa such as trypanosomes causing sleeping sickness have a simple flagellum, while others have a very complex flagellar arrangement as in *Trichonympha*, an organism inhabiting the guts of termites where it is responsible for the digestion of the cellulose in the wood eaten by the insects.

3 Ciliary movement in ciliates. These are usually covered over much of the cell surface with cilia which, although similar in structure to flagella, are shorter and have a co-ordinated motion so that waves of contraction pass over them.

(b) *Food* is characteristically taken up by the phagocytosis of solid particles such as bacteria. This digestion leads to the formation of food vacuoles in which digestion occurs, the indigestible material being liberated on the surface by evagination. In some protozoa such as amoeba there are no specialized areas on the surface, whereas ciliates usually have a 'mouth' for food uptake and an 'anus' for exit.

Many protozoa, especially parasitic forms, can feed by taking up soluble nutrients from their environment.

(c) Protozoa generally have no cell wall. However, some amoeboid forms have a solid structure which can best be described as a skeleton since it may be internal or external and has pores to allow for food uptake by phagocytosis. The marine foraminifera have complicated and multichambered calcium carbonate exoskeletons which can form geological deposits of chalk. Others (the radiolaria) have an endoskeleton of silica.

(d) The protozoa usually have a less complicated life cycle than algae or fungi, but there are some exceptions such as the parasitic protozoa (e.g. *Plasmodium*, the organism causing malaria).

(e) One of their most interesting aspects is the degree of specialization that can occur in a single cell. To take the well-known example of the unicellular *Paramecium*, there is a specialized food uptake system with gullet and mouth, an anus, contractile vacuoles, a co-ordinated array of cilia, and a complicated nuclear system containing both a micronucleus and a macronucleus. Organisms like *Paramecium* seem to represent the end-limit to which specialization of a single cell can go, the more fruitful evolutionary development being in the differentiation between the cells of multicellular organisms to form tissues.

3 The slime moulds

These are a group that traditionally have been studied by mycologists, because their most obvious attribute is the formation of macroscopic fruiting bodies. Their feeding phase, however, is amoeboid, and thus they are firmly protists. They live on the surface of decaying vegetation and often produce brightly coloured and ornate fruiting bodies. There are two main groups, the cellular slime moulds whose vegetative stage consists of single amoeboid cells which can aggregate to form a *pseudoplasmodium*, and the acellular slime moulds where a single amoeba produces a multi-nucleate *plasmodium* of indefinite size and shape which moves over the surface of a substratum engulfing food particles as it goes. The latter group represents the type of micro-organisms most beloved of the science fiction writer.

Let us consider the life history of a cellular slime mould, *Dictyostelium* (Fig. 3.4). The vegetative amoebae feed on particulate matter, especially bacteria; when the supply of food is exhausted, they aggregate into pseudoplasmodial groups, a process triggered by the localized secretion of cyclic-AMP which acts as a hormone-like substance and attracts cells together via a chemotactic response. The pseudoplasmodium moves towards the light as a single unit surrounded by a mucoid sheet secreted by the individual cells until eventually movement ceases. The multicellular structure then differentiates to produce a fruiting body. The front third of the cells form a tall thin stalk, strengthened by their

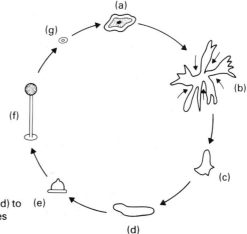

Fig. 3.4. The life cycle of a cellular slime mould. Vegetative amoeboid cells (a) aggregate together, (b) to form a pseudoplasmodium (c) which migrates (d) to form a fruiting body (e–f) which produces asexual spores (g).

49

forming cellulose walls, and then die. The rear two-thirds climb up this stalk, and form spores in a mass at the top. By this pattern of differentiation the original 100 000 or so individual amoebae dispersed through the detritus amongst which they were feeding have worked together to produce about 70 000 spores held up in the light and open air to be dispersed into the environment where they eventually germinate to start the process again. The similarity between this process and sporulation in myxobacteria (see earlier) is striking, and must be the result of convergent evolution.

4   The water moulds

These organisms, of two main types, the oomycetes and the chytrids, were originally placed in the fungal kingdom, and studied by mycologists, but are phylogenetically protists, most obviously because they form flagellated motile cells during their life histories. Among the oomycetes are many important plant pathogens, in particular *Phytophthora* which changed the course of history by causing the Irish potato famine of the 1840s.

The fungi

Let us consider some of the typical characteristics of filamentous fungi.

(a)   They are heterotrophic eukaryotic micro-organisms obtaining their food in a soluble form by uptake through the plasma membrane in a manner similar to that of prokaryotes.

(b)   They have a thick cell wall usually made of polysaccharides, nearly always with chitin microfibrils (see earlier).

(c)   They have no motile stages, and never form flagella.

(d)   They have a typically branched growth or *mycelium* made up of individual filaments called *hyphae*. The intertwined mycelia are produced by branching behind the hyphal tips during growth or by hyphal fusion; in some cases the aggregated hyphae may form large structures which bear a superficial resemblance to the tissues of higher plants (e.g. in mushrooms and toadstools).

(e)   Mycelia are often *coenocytic*—that is, they are composed of multinucleate tubes with the cytoplasm in continuous connection throughout. Some mycelia have no cross walls while others have septa with pores to allow cytoplasmic connection.

(f)   The majority are adapted to life in the soil where they are important in converting organic carbon to CO_2. The major

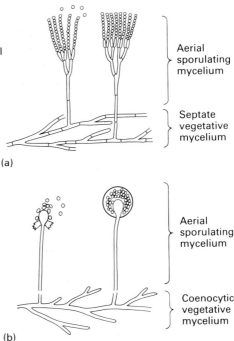

Fig. 3.5. Two types of asexual spore production and two types of vegetative mycelium. (a) *Penicillium* with asexual spores in as conidia; (b) *Mucor* with asexual spores in a sporangium.

Aerial sporulating mycelium

Septate vegetative mycelium

(a)

Aerial sporulating mycelium

Coenocytic vegetative mycelium

(b)

part of the organism (the *vegetative mycelium*) is concerned with the uptake of nutrients and attachment to a solid surface. Many others are pathogens, living in plant or animal tissues, or symbionts in lichens or plant roots (see Chapter 8).

(g)　From the vegetative mycelium specialized aerial hyphae are produced and from these asexual spores are differentiated (Fig. 3.5). The spores may arise as single cells at hyphal tips (*conidiospores* or *conidia*) or inside a structure called *sporangium* (*sporangiospores*). Sometimes spores are produced within a hyphae by a process analogous to endospore production in bacteria to produce resting spores (*chlamydospores*). While the vegetative mycelium is normally colourless, the aerial or reproductive mycelia are often brightly coloured. The whole process has a function of dispersal since each fungus produces enormous numbers of these light spores which are easily carried from one place to another by air currents.

(h)　Sexual spores (Fig. 3.6) can also be produced, as the result of sexual reproduction. A large variety of structures may be found to hold these sexual spores and the 'fruiting bodies' may be of considerable complexity. As a result of sexual reproduction, there is an alternation of haploid and diploid cells and this alternation may give rise to a complex life cycle (see Chapter 7). It is obvious that many types of spore-bearing structures can occur in fungi and great use is made of them in classification.

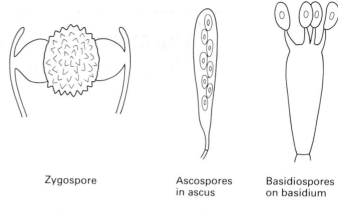

| Zygospore | Ascospores in ascus | Basidiospores on basidium |

Fig. 3.6. Different types of sexual spore in fungi.

Although the majority of fungi produce multicellular hyphae as indicated above, a few are characteristically unicellular. These are the *yeasts* which often show signs of the evolutionary origin by their ability under certain cultural conditions to grow as a mycelium. This phenomenon is known as mould:yeast *dimorphism*, and is especially important in human pathogenic fungi such as *Candida albicans*. It must be stressed that the term yeast has no taxonomic significance; it merely implies a unicellular growth form. There are yeast forms in all taxa of fungi. When the word 'yeast' is used in common parlance however, it refers to *Saccharomyces cerevisiae*, famed for its alcoholic fermentation (see Chapter 10).

The normal mode of yeast multiplication is by budding, but a few species divide into two equal daughter cells. Many also produce asexual and sexual spores.

4 : The Metabolism and Nutrition of Micro-organisms

One of the most important characteristics of micro-organisms is their high rate of growth. Many can have a generation time of under 30 minutes and this depends upon the ability to synthesize their own weight of cell material in this period. That is to say, they have a correspondingly high rate of metabolism. Indeed, there tends to be an inverse relationship between the rates of metabolism or growth and the size of a living organism. For example, if we take the rate of oxygen uptake per unit mass as a measure of energy production in an aerobic organism and if the figure for man is taken as unity, the following levels are obtained:

			Micro-organism	
Elephant	Man	Mouse	Eukaryote	Prokaryote
0.2	1.0	10	100	1 000

How is this high rate of microbial metabolism maintained? One of the main limiting factors is the rate of uptake of nutrients and removal of waste products, processes which both occur at the cell surface. The smaller the organism, the greater will be its ratio of surface area to volume, or to weight, and hence the easier it will be for it to have a high rate of metabolism and growth. This is illustrated by two examples:

(a) A 200 lb man
$$\frac{\text{Surface area}}{\text{Weight}} = \frac{24\,000 \text{ cm}^2}{10\,000 \text{ g}} = 2.4 \text{ cm}^2/\text{g},$$

(b) A typical bacterium

$$\frac{\text{Surface area}}{\text{Weight}} = \frac{1 \times 10^{-7} \, \text{cm}^2}{2 \times 10^{-12} \, \text{g}} = 50\,000 \, \text{cm}^2/\text{g}.$$

Of course, higher organisms have developed special mechanisms to increase their useful surface area such as intestines, lungs, kidneys and circulating bloodstreams. However, micro-organisms gain a considerable advantage from their small size and may actually increase their surface area still further by invagination of the cytoplasmic membrane (p. 21).

As a consequence of these characteristics, biologists have come to realize that micro-organisms are ideal for the study of some of the basic principles of biochemistry and molecular biology.

The chemical composition of micro-organisms

Before we consider microbial metabolism itself, it is essential to identify the chemical components which make up the microbial cell. Typical figures are shown in Table 4.1.

Table 4.1. The main components of typical microbial cells (figures as % of dry weight)

	Protein	Nucleic acid	Polysaccharide	Lipid	Peptidoglycan	Others
Prokaryote	50	10	10	10	10	10
Eukaryote	50	5	15	15	—	15
Monomer components of polymers	Amino acids	Nucleotides	Monosaccharides	—	Amino acids, Monosaccharides	—

There is a basic similarity in the composition of all living cells. At the same time, however, it must be realized that there are wide variations in chemical composition both between different groups of micro-organisms and between the same organism grown under different environmental conditions. A few examples of such variations are given below.

1 In all micro-organisms, the RNA content increases with growth rate (p. 23). Consequently, since prokaryotes in general grow faster than eukaryotes, this fact is reflected in their high levels of RNA.

2 Gram-negative bacteria have more lipid and less peptidoglycan than Gram-positive bacteria because of their different cell wall composition and structure (p. 24).

3 Cells grown in the presence of excess carbon source tend to have a high content of storage polymers such as lipid and polysaccharide (p. 23).

54

If we look at Table 4.1 it can be seen that 70–80% of the cell dry weight is made up of macromolecular polymers, the remainder being in the form of lipid and low-molecular weight substances such as metabolites, coenzymes and inorganic ions. A critically important fact to recognize is the invariability of most of these compounds in living organisms. The amino acids in proteins, the nucleotides in nucleic acids, many types of lipids and most coenzymes are of universal occurrence. Although there are considerable differences in the means employed by cells to obtain their energy, and the building blocks used for growth, as well as a broad range of biosynthetic ability, there tends to be a unity in the actual mechanisms of biosynthesis and in the final product in terms of cell composition.

The metabolism of a cell, then, must be directed to the synthesis of the following compounds:

20 L-amino acids,
2–4 peptidoglycan amino acids (in prokaryotes only),
5 purines and pyrimidines,
c. 10 monosaccharides,
c. 10 lipids,
c. 20 coenzymes and
A variety of other essential components.

The minimum number of organic compounds that must be either synthesized by the cell or provided in the environment as monomers or as cell constituents in their own right is probably between 100 and 200. In most micro-organisms, the majority of these constituents can be synthesized by the organism itself, the stages of this metabolism being as follows:

1   The provision of basic carbon intermediates, or building blocks, from external carbon sources. It is at this level that all living organisms are commonly divided into two major groupings—the heterotrophs and the autotrophs. In *heterotrophs* the sources of carbon and energy are generally the same organic substances which are broken down partly to supply building blocks and partly to provide energy. In *autotrophs* the carbon source is CO_2. Energy is consequently not available from the catabolic metabolism of organic substances and must therefore be provided by another mechanism; for example, photosynthesis by *phototrophs*, or the oxidation of inorganic compounds by *chemotrophs*. Although the distinction is clear enough for plants (autotrophs) and for animals (heterotrophs), for micro-organisms the borderlines between the various nutritional groups can be blurred with many showing environment-dependent variation in the determinative characteristics.

2   The provision of energy by the formation of ATP, generally from ADP or occasionally from AMP. (For simplicity, only ADP will be mentioned in this book.)

3   The provision of reducing equivalents in the form of the reduced pyridine nucleotides; either NADH or NADPH, and referred to here as NAD(P)H. In heterotrophs NAD(P)H is generated along with ATP during the catabolism of the organic carbon sources. In autotrophs the provision of NAD(P)H parallels that of ATP, arising from the photolytic

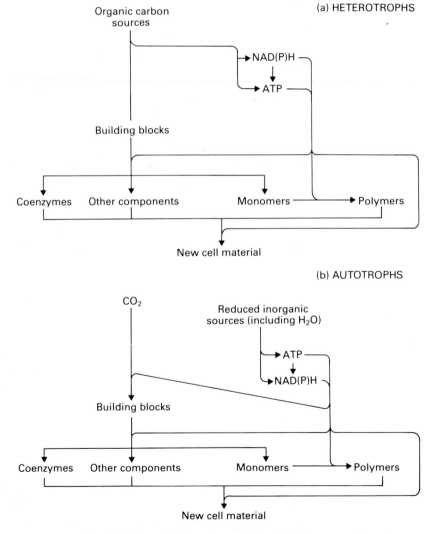

Fig. 4.1. The major steps in metabolism of (a) heterotrophs, and (b) autotrophs.

splitting of water in plant and algal photosynthesis, and from the oxidation of reduced inorganic compounds in bacterial photo- and chemo-autotrophs.

4   The conversion of these building blocks into the monomers, coenzymes and other cell components noted above.

5   The polymerization of the monomers.

6   The formation of the quaternary structure of the polymers, together with other essential components, and their movement to the appropriate part of the cell where they form the structures necessary for their normal functioning. These processes are summarized in Fig. 4.1.

We can now consider the various stages of metabolism beginning with *anabolism*, the largely universal sequence of pathways leading from the carbon building blocks to the complete cellular structure in a series of ATP and NAD(P)H–requiring reactions. It is, however, in *catabolism*, which encompasses all the mechanisms for the generation of the central carbon intermediates and for NAD(P)H and ATP production, that the striking variation and versatility of the micro-organisms becomes most evident.

The biosynthesis of monomers and coenzymes

It is possible to identify a number of central metabolic intermediates which constitute the basic building blocks from which the cellular carbon is synthesized: glucose, ribose, glycerol, phosphoenolpyruvate (PEP), pyruvate, acetyl coenzyme-A, α-oxoglutarate (OGA), oxaloacetate (OAA), etc. Leading from these metabolic intermediates is a series of specific pathways concerned in the biosynthesis of the monomers, coenzymes and other essential compounds needed for growth. Although the pathways themselves are largely ubiquitous in nature, there is a considerable variation amongst micro-organisms in the range of biosynthetic pathways present. While some have a complete set and can synthesize all their organic components for themselves, others have a restricted range of biosynthetic ability.

The essential monomers must eventually be produced in an activated form so that no additional energy source is required for their polymerization. Their synthesis therefore also requires the input of energy and reducing equivalents in the form of ATP and NAD(P)H, respectively, to facilitate the carbon transformation. These activated monomers (e.g. amino acyl–tRNA, nucleoside diphosphate sugars, nucleoside triphosphates, etc.) are usually the same in micro-organisms as in higher organisms.

57

Polymerization

In general, the methods used in polymerization are similar in all living organisms, differing only in the type of polymer produced.

Homopolymers

e.g. –A–A–A–A–A

Here an identical monomer is repeated along the chain. Polymerization requires a single polymerase acting on the activated monomer (A^*), a reaction usually also requiring some polymer $[(A)_n]$ as a primer,

i.e. $(A)_n + A^* \rightarrow (A)_{n+1}$

Consider two typical examples in microbiology:
(a) Polyglucoses

$$(glucose)_n + ADPG \rightarrow (glucose)_{n+1} + ADP$$

where $(glucose)_n$ = glycogen, and ADPG = adenosine diphosphate-glucose,
and

$$(glucose)_n + glucose\text{-}fructose \rightarrow (glucose)_{n+1} + fructose$$

where $(glucose)_n$ = dextran, and glucose-fructose = sucrose.
(b) Poly-β-hydroxybutyrate

$$(\beta\text{-hydroxybutyrate})_n + \beta\text{-hydroxybutyryl CoA}$$
$$\rightarrow (\beta\text{-hydroxybutyrate})_{n+1} + CoA$$

Heteropolymers with a repeating unit (regular heteropolymers)

e.g. –A–B–C–A–B–C–A–B–C– or $(A–B–C)_n$

Here more than one monomer occurs but the polymer is built up from repeating units of varying degrees of complexity. Polymerization usually occurs by a preliminary biosynthesis of the activated repeated unit followed by its polymerization:

i.e. $A^* + B^* \rightarrow AB^*$
$AB^* + C^* \rightarrow ABC^*$
$(ABC)_n + ABC^* \rightarrow (ABC)_{n+1}$

In this example, the specific nature of the polymer is determined by the specificity of three enzymes. Typical microbial polymers formed in this way are the heteropolysaccharides in the cell wall and capsule, and the prokaryotic peptidoglycan.

58

Heteropolymers without a repeating unit (irregular heteropolymers)

e.g. –A–D–A–C–B–D–A–A–

Such heteropolymers exist in nature in the form of nucleic acids and proteins. In contrast to the previous examples where the nature of the polymer is determined by enzyme specificity, here it is determined by a coding or template mechanism (see Chapter 7). Both the general methods employed in transcription and translation, and the actual code used, appear to be common to all living organisms. Their basis in the double-helix of DNA has almost passed into folk lore and will be assumed to be known to the reader of this book.

Assimilation of components into the structural and functional integrity of the cell

Many of the chemical components of the cell are produced in a fully active state at the site where they are required to fulfil their role. Thus, soluble enzymes are formed on ribosomes and are released directly into the cytoplasm, whilst many cell wall macromolecules are polymerized at specific sites on the outside surface of the cytoplasmic membrane. Here the activated repeating units are joined to the cell wall *in situ* by the action of enzymes bound into the cytoplasmic membranes. On the other hand, other components require transportation and incorporation into appropriate structures before they become fully active. Thus, nutrient transport carrier and electron-transport proteins are incorporated into the cytoplasmic membrane; DNA and RNA polymerases into the nucleus (in eukaryotes), ribosomal proteins into ribosomes and so on. The study of the mechanisms involved in the recognition and location of protein molecules within membranes and organelles is currently a particularly intriguing and important area of research activity in cell biology.

The production of carbon intermediates, energy and reducing equivalents

In discussing the growth and metabolism of micro-organisms it is important to consider the flux of carbon compounds in direct association with the production and utilization of ATP and NAD(P)H. We have already noted their interdependence in the biosynthesis of the monomeric and polymeric constituents of cells, and the same is true for the pathways

59

concerned with provision of the necessary building blocks from which these cellular components are ultimately synthesized.

Heterotrophs

In heterotrophs the carbon sources are catabolized by a series of reactions to provide a group of essential metabolites which are used as the basic building blocks for the synthesis of the monomers, coenzymes and fundamental structural units of the cell. Since these catabolic reactions are essentially oxidative in character they also give rise to the production of ATP and NAD(P)H. Thus, enzymes catalysing inter-conversions such as those of the glycolytic pathway and the tricarboxylic acid cycle are generally common to all cellular life; we shall refer to them collectively as consituting the *basal intermediary metabolism*. An important function of these enzymes is that whereas in heterotrophs they are concerned both with the production of energy and reducing equivalents, and with the synthesis of the essential cell building blocks, in autotrophs where CO_2 is the carbon source for growth they usually serve an anabolic or biosynthetic role. Such enzymes and pathways which can function in both catabolism and anabolism are known as *amphibolic* and are usually present irrespective of the environmental conditions. In other words, they are *constitutive*.

Leading into these basal pathways are the enzymes concerned in the specific utilization of particular carbon sources. Since many heterotrophic micro-organisms are capable of using over a hundred different carbon sources and since many specific enzymes may be involved in each of the pathways, a cell has the potential to produce a very large number of catabolic enzymes. In general, such enzymes are only formed when they are actually required, that is when the specific carbon source is present in the environment. These are *inducible enzymes* and their synthesis requires the presence of an inducer which is usually the particular carbon source. An example is the induction of the enzyme α-glucosidase (maltase) by the inducer maltose. As a result, maltose is hydrolysed to glucose which can be broken down by constitutive enzymes.

During the chemical breakdown of the organic carbon and energy source, heterotrophs generate ATP by two general phosphorylation mechanisms. In *substrate-level phosphorylation* the organic substance undergoes an energy-yielding, or exergonic reaction in such a way that a phosphate derivative is formed which has sufficient free energy to transfer its phosphoryl group directly to ADP to produce ATP. In *oxidative phosphorylation* an electron donor (AH_2) is oxidized and the electrons (usually paired) so produced pass through an electron-transport system to a terminal electron acceptor (B) in such a way that the energy released is partly utilized to produce ATP from ADP (Fig. 4.2).

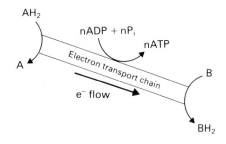

Fig. 4.2. The concept of oxidative phosphorylation.

or $AH_2 + B + nADP + nP_i \rightarrow A + BH_2 + nATP$

n, which represents the number of energy-rich phosphate bonds produced in the passage of a pair of electrons through the electron-transport system, may have values between 1 and 4. The figure depends ultimately on the energy released in the oxidation of AH_2 at the expense of B, and therefore on the difference in the redox potential of the $AH_2 \rightarrow A$ system and the $B \rightarrow BH_2$ system. The greater this difference, the more ATP that can theoretically be produced.

Consider the breakdown of glucose. Most micro-organisms employ the glycolytic pathway for the initial attack, with the resultant production of two molecules of pyruvic acid, two molecules of NADH and two molecules of ATP generated by substrate-level phosphorylation:

$$\text{i.e. } C_6H_{12}O_6 + 2NAD^+ + 2ADP + 2P_i$$
$$\rightarrow 2CH_2CO.COOH + 2NADH + 2H^+ + 2ATP$$

For the reaction to continue, the reduced pyridine nucleotide must be reoxidized, a process that may produce additional useful energy.

There are indeed three general mechanisms whereby NAD(P)H, produced during glycolysis or from any dehydrogenation reaction in other catabolic sequences, can be reoxidized:

(a) NAD(P)H can donate its reducing equivalents to the electon transport chain in aerobic or anaerobic respiration, with coupled ATP synthesis.

(b) NAD(P)H can donate its reducing equivalents to an intermediate metabolite to give rise to a reduced fermentation product.

(c) NAD(P)H can be used as the source of reducing equivalents during reductive biosynthesis.

Three general methods exist by which a carbon and energy source can be broken down to provide energy:

(a) *Aerobic respiration.* In aerobic respiration, the carbon and energy source is broken down by a series of reactions, the oxidation stages occurring at the expense of oxygen as the terminal electron acceptor. By far the greater amount of useful energy produced during aerobic respiration comes from oxidative phosphorylation. The complexity of the electron transport system varies amongst different micro-organisms. Thus, yeasts can have a series of carriers in their mitochondria very similar to those of higher animals, while bacteria often have fewer cytochromes and branched electron transport systems which frequently possess smaller numbers of phosphorylation stages.

In the classical example of the complete oxidation of glucose to carbon dioxide, the pathways involved appear to be common to microbes, plants and animals; glucose is first converted by the glycolytic pathway to pyruvic acid which is then oxidized to carbon dioxide through the tricarboxylic acid cycle, the major part of the ATP formed being by oxidative phosphorylation. Assuming three energy-rich phosphate bonds are formed during each passage through the electron-transport chain, the overall oxidation can be explained as follows:

$$C_6H_{12}O_6 + 6O_2 + 38ADP + 38P_i \rightarrow$$
$$6CO_2 + 6H_2O + 38ATP$$

Of the 38ATP formed, only four result from substrate-level phosphorylation.

Although carbon dioxide is the commonest end-product of the oxidation of a carbon and energy source, some micro-organisms carry out only a partial oxidation. For example, some fungi produce oxalic acid

$$C_6H_{12}O_6 + 4.5O_2 \rightarrow 3(COOH)_2 + 3H_2O$$

(b) *Anaerobic respiration.* Although oxygen is the most common and efficient electron acceptor, some prokaryotes are able to utilize certain inorganic compounds as alternative terminal electron acceptors in a process called anaerobic respiration (Fig. 4.3). Thus, nitrate is reduced to nitrite, nitrous oxide or molecular nitrogen, and sulphate is reduced to sulphide. In a parallel series of reactions in methanogenic bacteria, carbon dioxide is reduced to methane during the oxidation of hydrogen; although carbon dioxide is acting as a

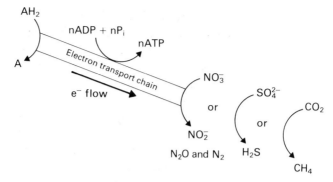

Fig. 4.3. Anaerobic respiration.

terminal electron acceptor, the mechanism does not involve a conventional electron transport chain.

The following points should be noted:

1   The pathways for the breakdown of a carbon and energy source in aerobic and anaerobic respiration are the same, although generally somewhat restricted in the latter. The only significant difference resides in the fate of the electrons produced in the oxidation steps.

2   The amount of ATP formed in the passage of two electrons through the electron transport chain depends on the difference in redox potential between the electron donor and the electron acceptor. Since these alternative inorganic electron acceptors have lower redox potentials than oxygen, less ATP will usually be produced in anaerobic compared with aerobic respiration.

3   Anaerobic respiration plays an important part in maintaining the cycle of elements in natural environments (see Chapter 9).

(c)   *Fermentation.* In fermentation* no external electron acceptor is required. Instead the carbon and energy source is broken down by a series of reactions which produce ATP by substrate-level phosphorylation. Although oxido-reductions, or redox reactions occur, they must be balanced so that the average state of oxidation of the products is the same as the substrate.

---

*The term 'fermentation' is often used loosely to describe large-scale cultivation of micro-organisms in industry. This is an incorrect usage of the term as such processes are generally aerobic and involve complete oxidation of the carbon source.

There is a wide range of fermentation products from glucose but only two will be mentioned here. The lactic acid bacteria reduce pyruvic acid directly to lactic acid:

$$2CH_3CO.COOH + 2NADH + 2H^+ \rightarrow$$
$$2CH_3CHOH.COOH + 2NAD^+$$

The end result is as follows:

$$C_6H_{12}O_6 + 2ADP + 2P_i \rightarrow 2CH_3CHOH.COOH + 2ATP$$

Yeasts, on the other hand, carry out the slightly more complex alcoholic fermentation involving the further breakdown of the pyruvate fermentation:

$$\text{i.e. } 2CH_3CO.COOH \rightarrow 2CH_3CHO + 2CO_2$$

$$2CH_3CHO + 2NADH + 2H^+ \rightarrow$$
$$2CH_3CH_2OH + 2NAD^+$$

The end result is the formation of ethanol and carbon dioxide:

$$C_6H_{12}O_6 + 2ADP + 2P_i \rightarrow 2CH_3CH_2OH + 2CO_2 + 2ATP$$

These are only two of the many end-products of fermentation in micro-organisms. The nature of these end-products is often an important factor in classification and many are also of great industrial value. Anaerobic growth by fermentation is much less efficient energetically than aerobic growth, the yield of cells from a unit amount of glucose usually being about a tenth. This is less than the difference in the ATP yield per molecule of glucose oxidized or fermented since a higher percentage of glucose will be assimilated to cell material in aerobic growth as compared to anaerobic growth.

Autotrophs

In autotrophs CO_2 is the source of carbon. It is converted into the basal intermediary metabolic system by a series of reactions (the Calvin cycle) which seem to be common to all such organisms, be they photoautotrophs or chemoautotrophs. Unlike the oxidative catabolism of heterotrophs, however, the Calvin cycle is reductive and energy-consuming, or endergonic, and thus requires an input of reducing power and energy in the form of NAD(P)H and ATP respectively.

Autotrophs fall into two groups according to their mode of energy production:

1 *Photoautotrophs.* The photoautotroph heterotrophs, energy by photophosphorylation, a process involving the excitation of chlorophyll molecules by light with the con-

sequent emission of electrons which are transferred in a reductive reaction to an acceptor of low redox potential. This primary acceptor can then be reoxidized by transfer of the electrons through an electron-transport system with concomitant production of ATP by a process analogous to that of oxidative phosphorylation (Fig. 4.4).

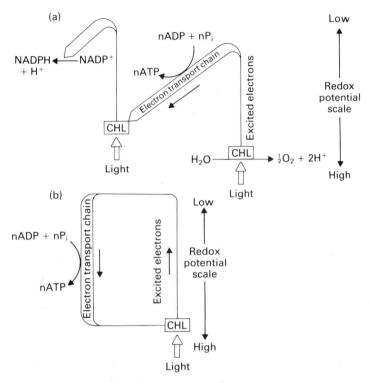

Fig. 4.4. A simplified scheme for (a) non-cyclic and (b) cyclic photophosphorylation (arrow shows direction of electron flow).

In *non-cyclic photophosphorylation* the electrons lost from the light-excited chlorophyll are replaced from the photolytic splitting of water with the consequent production of oxygen. The electrons from the oxidative ATP-synthesizing arm are donated to a second chlorophyll reaction centre which on excitation donates electrons to a second primary acceptor, generally of lower redox potential than the acceptor associated with the first chlorophyll reaction centre. From this second acceptor electrons are transferred through an abbreviated electron-transport chain to give NAD(P)H. This oxygenic non-cyclic photophosphorylation which is found in plants, algae and in blue-green algae, or cyanobacteria, thus generates ATP from light energy and NAD(P)H from water (Fig. 4.4).

65

Eubacterial photophosphorylation is generaly *cyclic* and does not involve the splitting of water with consequent oxygen production. The reaction centre contains bacterio-chlorophylls and other light-gathering pigments and they absorb light of higher wavelengths. Since the excited electrons are returned to the chlorophyll after the ATP–synthesizing oxidative arm, NAD(P)H is not a light-dependent product of bacterial cyclic photophosphorylation (Fig. 4.4).

In certain cases a few simple organic compounds or hydrogen can supply reducing equivalents, but most commonly the electron donor is hydrogen sulphide or elemental sulphur. Although the reaction is light-independent, reduced sulphur compounds can only donate electrons at the redox level of components more oxidized than NAD(P) and thus the reduced form NAD(P)H can only be generated by *energy (ATP)–driven reversed electron transport* (Fig. 4.5).

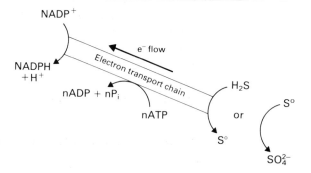

Fig. 4.5. Reversed electron flow found in most autotrophic bacteria. The example shown is for *Thiobacillus*, an organism capable of oxidizing many reduced sulphur compounds.

2   *Chemoautotrophs.* These bacteria obtain their energy by the oxidation of inorganic substrates, usually at the expense of oxygen as the terminal electron acceptor. ATP is formed by oxidative phosphorylation and the basic difference from energy production in aerobic heterotrophs is that the electron donor happens to be inorganic. Some examples of these intriguing organisms are shown in Table 4.2.

The chemoautotrophs are an important group from an economic point of view. One of them has also become something of a 'cause célèbre'. *Thiobacillus thioxidans* was used to test the idea of the unity of general biochemical principles since it was considered to be as bizarre a living organism as existed. It gets its carbon from CO_2, its energy by oxidizing inorganic sulphur compounds to sulphuric acid, it will grow at a very acidic pH and it cannot grow on normal

Table 4.2. The electron donors and their products in the energy-producing reactions of some chemoautotrophic bacteria

	Electron donor and product	
Nitrifying bacteria	NH_3	NO_2^-
	NO_2^-	NO_3^-
Colourless sulphur bacteria	H_2S	S^0
	S^0	SO_4^{2-}
Hydrogen bacteria	H_2	H_2O
Methane bacteria	CH_4	CO_2
Iron bacteria	Fe^{2+}	Fe^{3+}

organic substrates such as glucose which may actually be inhibitory. However, we have seen that there is nothing unusual in the actual mechanism of ATP production; the electron donor for oxidative phosphorylation just happens to be an inorganic sulphur compound. The intermediary metabolism of the organism is little different from that of any other autotroph which uses CO_2 as the sole carbon source.

As with the phototrophic sulphur bacteria, all of these electron donors (with the exception of hydrogen) couple with the electron-transport chain at a point close to the oxidized end. This accounts both for the low ATP yield, and the need for NAD(P)H production by ATP–driven reversed electron transport in these organisms (Fig. 4.5).

The nutrition of micro-organisms

We can see then that the astonishing diversity of micro-organisms relates primarily to the various alternate mechanisms utilized in producing the carbon intermediates, energy and reducing equivalents required for cellular biosynthesis and growth. It is these patterns of catabolism that determine the cells' basic nutritional requirements; the grouping of micro-organisms as autotrophs or heterotrophs, as phototrophs or chemotrophs is essentially a nutrition-based classification.

There are, however, 100–200 monomers, coenzymes and other essential chemical components required for the growth of micro-organisms. These substances must either be synthesized by the cell itself or be supplied in the environment as nutrients. In other words, there is an inverse relationship between the biosynthetic capabilities of a micro-organism and its nutritional requirements. Some cells are capable of synthesizing all their cellular components from the basic metabolic building blocks and require only very simple growth media. Others can synthesize very few of their components and therefore have complex nutritional requirements. This range can be illustrated in the provision of organic nitrogenous compounds. The most important of these from the point of view of bulk requirements are the amino acids and nucleotides needed for protein and nucleic acid synthesis. Some micro-organisms have the ability to utilize molecular nitrogen by a reductive process called *nitrogen fixation*, leading to the production of ammonia which in turn is converted into organic nitrogen.

$$N_2 \xrightarrow[\text{nitrogen fixation}]{} NH_3 \longrightarrow \quad \text{Amino acids and nucleotides}$$

Such nitrogen-fixing micro-organisms are therefore capable of utilizing atmospheric nitrogen as the sole nitrogen source; others require a fixed inorganic form of nitrogen such as ammonia which can be converted into all the organic forms required; still others are incapable of the biosynthesis of most of all of their amino acids and nucleotides and these require their nitrogen in a complex organic form.

Consider micro-organisms with simple growth requirements. The main cellular elements are carbon, hydrogen, oxygen, nitrogen, sulphur and phosphorus. In a heterotroph, carbon is provided by suitable organic substances, hydrogen and oxygen are provided by water, whilst nitrogen, sulphur and phosphorus can be provided as inorganic ions such as ammonium, nitrate, sulphate or phosphate. Such organisms can be grown in a *synthetic or defined medium* in which typical components might be glucose, NH_4Cl, Na_2SO_4, K_2HPO_4, $MgSO_4$ and $FeSO_4$. The other essential elements for microbial growth (e.g. Mn, Ca, Co, Mo, Cu, Zn) are usually present in sufficient quantities as contaminants of the above components. For the growth of autotrophs, CO_2 is added as the carbon source. An appropriate electron donor must also be provided and, in photoautotrophs, the cells must be illuminated.

A series of more complicated *synthetic media* can be made up to contain the additional nutrients required for a particular micro-organism. For example, *Salmonella typhi* (the organism causing typhoid fever) requires the addition of only the amino acid tryptophan to a simple synthetic medium as it is unable to synthesize this particular amino acid for itself. Other micro-organisms have much more complicated needs. Thus, *Leuconostoc mesenteroides* requires in addition to the basic medium, acetate, nineteen amino acids, four purines and pyrimidines and ten coenzymes (which, when required as nutrients, are usually called microbial vitamins, or growth factors); it obviously has a very restricted biosynthetic capability. These complex synthetic media are very expensive and tedious to make and their use is largely restricted to research purposes. In practice in the routine laboratory what is usually needed is a medium that will support the growth of as many micro-organisms as possible, is reproducible in content and is cheap and easy to produce. A variety of such general laboratory media are manufactured, but most contain two main nutrient sources:

1   A *protein hydrolysate* (sometimes called peptone) to provide carbon, energy, nitrogen and amino acids.

2   A *natural extract* to provide other essential nutrients such as vitamins and inorganic salts. Meat or yeast extract are commonly employed but may, in addition, be supplemented by blood, serum or egg. An example of such a general laboratory medium is nutrient broth which usually contains 1–2% peptone, 1% meat extract and additional NaCl to bring the osmotic pressure up to the optimum for most bacteria.

Microbiological assay

The realization of the range of substances required for microbial growth, and of the specificity of these requirements, led to the development of assay methods for these substances using micro-organisms. If the total amount of microbial growth in batch culture is measured in a series of media containing variable amounts of a particular essential nutrient, results similar to those in Fig. 4.6 are obtained. At lower levels of the essential nutrient there is a linear relationship between its concentration and the amount of growth. Above this region, the total amount of growth becomes

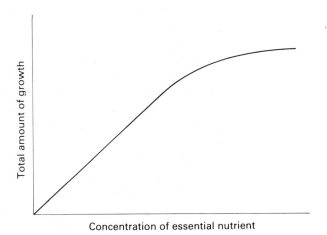

Concentration of essential nutrient

Fig. 4.6. Microbiological assay. The relationship between growth and the concentration of an essential nutrient.

constant as some factor other than the level of the chosen essential nutrient limits growth. The linear part of the curve can then be used to assay the nutrient. In theory, any substance required specifically for the growth of a micro-organism can be assayed but the method has been particularly valuable for vitamins since they can be measured with a greater specificity and sensitivity than by existing chemical methods.

Membrane transport and the uptake of nutrients

In order for nutrients to supply the cells' metabolic needs for growth, they must first cross the cytoplasmic membrane and enter the cell. In the case of gases such as oxygen this is done by simple diffusion, but special transport mechanisms are required for other nutrients since the membrane is largely impermeable to materials such as sugars, amino acids etc., and ions such as potassium or sulphate.

There are two broad mechanisms employed by micro-organisms for the uptake of particulate matter and large polymers that are themselves unable to pass through the membrane. Certain protozoa can engulf the material by the process of endocytosis (p. 48) so that it enters the cell within a membrane-bounded vesicle which can then fuse with a second lysosome vesicle containing hydrolytic enzymes capable of degrading the particulate or polymeric material. Fungi and bacteria are incapable of endocytosis as a consequence of their cell wall structure external to the membrane, and they hydrolyse polymers extracellularly through the secretion of protease, nuclease and saccharase enzymes.

Transport, or translocation of soluble ions and molecules across the cytoplasmic membrane is achieved through the action of transport carrier mechanisms, sometimes referred to as permease or porters. These are protein in nature and are coupled to cell metabolism. They demonstrate two striking features: they show specificity, and they can accumulate nutrients within the cells to a higher concentration than that at which they are present in the surrounding medium. The first of these allows control over what enters the cell, and the second is of crucial importance where the micro-organism is growing in a dilute nutrient environment, as is often the case.

It is interesting to note that these uptake mechanisms have their direct counterparts in the exit of metabolic products, and the deposition of cell walls and capsules, and of extra-cellular enzymes.

Regulation and integration of metabolism

So far in this chapter we have considered the individual reactions and pathways of anabolism and catabolism. The viability and growth of the cell, however, depends crucially on the regulation and integration of these separate metabolic activities. This aspect of cell physiology embraces major components of biochemistry and genetics.

As a rule of thumb it can be said that most organisms have control mechanisms which minimize the unnecessary expenditure of carbon and energy on processes which are not necessary under the prevailing conditions of growth. A simple example of this regulation of metabolism is the effect of provision of a cell monomer in the medium. If an amino acid is supplied to the organism, then the further synthesis of that amino acid by the cell is switched off. This is achieved by two types of mechanism: control of enzyme activity and control of enzyme synthesis. These two mechanisms together can account for all the regulatory phenomena of cells, although the molecular details will vary from one metabolic system to another.

(a) *Control of enzyme activity*. Frequently it is found that the enzymes involved in the first step of the synthesis of monomers are subject to feedback inhibition by the product(s) of the pathway. This negative feedback is common to most anabolic pathways and is a means of integrating supply and demand. In other cases the products of enzymes early in a pathway can have a stimulatory effect on the later enzymes, but this is quite rare and is usually involved with the regulation of energy production in catabolic sequences.

(b) *Control of enzyme synthesis*. The molecular mechanisms involved in this phenomenon are described later (Chapter 7). Suffice to say that in general enzymes are not produced unless required for the biosynthesis of a monomer or for the degradation of a carbon source. However, there are subtleties in that organisms have distinct preferences for the carbon source they will utilize. This phenomenon has been best investigated in the enteric bacteria and is termed catabolite repression. In these organisms glucose is the preferred carbon source and the synthesis of the necessary enzymes and the consequent utilization of all other carbon sources is prevented as long as glucose is available. The mechanism has been elucidated and appears to involve two processes, the prevention of inducer entry into the cells by inhibition of specific transport systems and the regulation of the synthesis of a small molecule called cyclic adenosine monophosphate (usually abbreviated to cAMP). The concentration of cAMP in the cell reflects the energy status of the cell; it is low when energy is freely available, for example in the presence of a rapidly metabolizable carbon source such as glucose. Thus, cAMP acts as a reporter molecule for the energy status of the cell and is used to control the expression of a large number of genes.

It is now known that the integrated functioning of the cell depends on monitoring the concentrations both of the individual products of enzyme pathways and of the special reporter molecules, often called 'alarmones'. These have evolved to 'report' the status of the cell's environment by monitoring the availability of readily utilizable carbon and nitrogen sources, the rate of protein synthesis, and the concentration of potentially lethal agents.

By these various mechanisms the separate components of the cell's metabolism operate at optimum efficiency and are orchestrated to satisfy the overall requirements of cell growth.

5 : The Growth of Micro-organisms

Growth may be defined as the orderly increase in all cellular constituents and results from the biosynthetic and energy generating processes described in the previous chapter. As such, it may refer to the increase in mass of a single cell or to the increase in size of a population of cells. This chapter considers both these levels of growth of unicellular micro-organisms and also growth of fungi and actinomycetes, which have a multicellular, mycelial growth form.

The cell cycle

Unicellular bacteria typically undergo binary fission, in which a cell grows to twice its size and then divides into two identical daughter cells, each of which grows in the manner of the parent cell until it too divides. The sum of all the processes occurring between cell divisions make up what is called the *cell cycle*. To function efficiently, the cell cycle must involve the controlled regulation and timing of biosynthesis, replication of DNA, equal partitioning of nuclear material, septation and separation of daughter cells.

To appreciate the enormity of this task we may consider growth of an *Escherichia coli* cell at 37°C in a rich nutrient medium. We have seen (p. 22) that the chromosome in *E. coli* consists of a circular double stranded DNA molecule with a circumference of approximately 1 mm. Replication of the chromosome is initated at a point called the origin and proceeds bidirectionally at two replication forks which meet at a point called the terminus (Fig. 5.1). At each replication fork the double stranded DNA molecule is split and nucleotides are added to each naked single strand. In *E. coli* growing at 37°C replication of this length of DNA, almost 1000× the

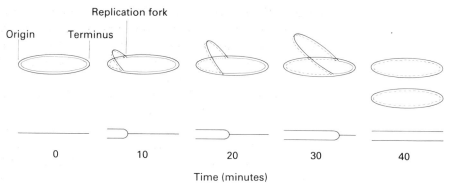

Origin   Terminus

Replication fork

Time (minutes)

Fig. 5.1. Replication of a circular double-stranded DNA molecule begins at a point called the origin and proceeds via two replication forks which travel around the chromosome to meet at the terminus. A simplified representation of chromosome structure during replication is also shown.

cell length, takes only 40 minutes. This part of the cell is called the C−period.

Remarkably, under these conditions cells will divide, on average, every 20 minutes. Following each C−period there is

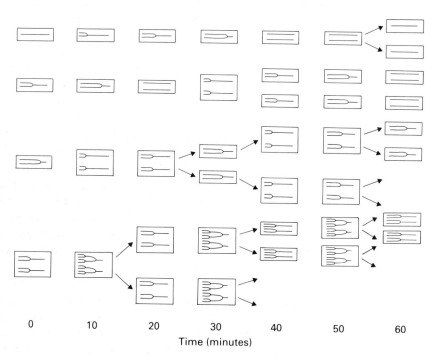

Time (minutes)

Fig. 5.2. The regulation of chromosome replication and cell division in *E. coli* growing with doubling times of (a) 60 minutes; (b) 40 minutes; (c) 30 minutes and (d) 20 minutes.

74

a gap of 20 minutes (called the D–period) between termination of replication and subsequent septation, which physically separates the daughter cells. Each cell must therefore plan ahead and initiate rounds of replication which will be completed in its granddaughter cells.

To illustrate how DNA replication and cell division are linked we can refer to Fig. 5.2 and consider first a cell growing in a relatively poor medium such that the time between divisions, the doubling time, is 60 minutes. A new-born cell will contain a complete chromosome which immediately begins replication. After 40 minutes (C–period) the cell will contain two complete chromosomes and after a further 20 minutes (D–period) the cell will divide, segregating the two chromosomes. The combined length of the C– and D–periods equals 60 minutes; there are no gaps and the cell cycle is said to operate continuously.

If the cell is growing in a richer medium which allows a doubling time of 40 minutes then the C– and D–periods exceed the doubling time and cell cycles must overlap, i.e. the D–period of one cycle will overlap with the first 20 minutes of the C–period of the next cycle. Consequently, a new-born cell must contain a chromosome which is already half-replicated. After 20 minutes it will contain two chromosomes, each of which will begin a new round of replication. When cell division occurs 20 minutes later, each daughter cell will again contain one half replicated chromosome.

A new-born cell in medium giving a doubling time of 30 minutes will contain a chromosome which is three quarters replicated. The cycles must now overlap by 30 minutes and a second round of replication will be initiated before the previous round is completed. This means that for a period of ten minutes the chromosome will have six replication forks, two from the first round and four from the second round of replication. The overlap is even greater in a cell dividing every 20 minutes and a new-born cell must contain two half replicated chromosomes, each of which initiates a further round of replication. The precise mechanisms by which the cell integrates these processes and partitions nuclear material have still to be elucidated.

New cell wall material may be added diffusely, over large regions of the cell surface, or in localized regions such as the cell poles. The most detailed studies have been carried out on *Streptococcus faecium* where wall growth is localized and is intimately linked to septum formation. Each cell possesses a wall band which splits to form a septum which then grows inwards while the external surface splits to develop into a new pole. Eventually the inner edges meet to form a complete septum, by which time, of course, nuclear material must have replicated and segregated so that each daughter cell contains at least one complete chromosome.

Although we often think of prokaryotic micro-organisms as having a rather restricted range of forms in comparison with higher organisms, there is in fact much variety, and many unicells do not undergo symmetric division into identical cells but grow by budding. This results from the existence of obligate polar cell growth. For example, in one such bacterium, *Hyphomicrobium*, a filament or stalk grows, and then develops a new daughter cell (see p. 41). The two cells become separated by formation of a plug or septum in the filament. We therefore have distinct mother and daughter cells. Since it is thought that a mother can only produce a limited number of daughters, we must introduce the concept of mortality which is inapplicable to unicellular organisms undergoing binary fission. This mode of growth allows for a greater degree of morphogenesis and a consequent greater variety of form. One of the most complex prokaryotic cell cycles is that of the budding bacterium *Rhodomicrobium vanellii*. It exists in one form as flagellated swarm cells which, with sufficient illumination, shed their flagella and develop into ovoid cells. Each of these then produces a filament from which a daughter cell is formed by budding. A maximum of four daughter cells can be produced in this way, each of which can itself produce daughters. Under certain conditions cells remain attached by the filaments producing chains of cells, often with branch points. Alternatively, the daughter cells may develop into motile swarm cells or into non-motile angular exospores which themselves will eventually grow by budding. Prokaryotic micro-organisms therefore have the potential for morphogenesis and it is likely that many interesting and varied organisms are still to be discovered as

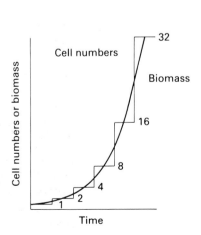

Fig. 5.3. The growth and multiplication of a single cell.

we improve our techniques of isolation from natural environments.

We have concentrated here on the growth of prokaryotic cells. The eukaryotic cell cycle is usually divided into a division phase (M) and a synthesis phase (S), separated by two 'gap' phases (G_1 and G_2) when DNA is not synthesized. Chromosome replication proceeds bidirectionally from several, rather than a single site and cytoskeletal structures are involved in separation of daughter chromosomes. Such structures are also involved in cell division and cell growth of eukaryotes and help to provide the greater range of cellular morphologies seen in these organisms.

Population growth

We can consider the growth of a population of unicells as an increase in total cell mass (*biomass*) resulting from growth, or as an increase in total cell number resulting from both growth and division. This can be seen in the growth and multiplication of a single cell (Fig. 5.3) and the same results occur in a synchronously dividing population of cells (see p. 86). However, the microbiologist is normally dealing with a large number of cells; for example, a bacterial colony on an agar plate contains about 10^8 cells, a suspension of 10^8 cells/ml will be only just visibly turbid, whilst 1 mg dried bacteria contains about 3×10^8 cells. With such large populations, the chances are that the percentage of cells dividing in any one small period of time will be the same as that in any other period and if we plot biomass or cell number against time, we will get the same curve. In other words, the cells in the culture are dividing asynchronously. Consequently, we can normally consider growth and multiplication as being synonymous and we can measure whichever is most practicable.

In order to study multiplication, the actual number of organisms must be measured. The straightforward way to do this is to use a counting chamber under the microscope, but this simple process immediately brings up a problem. How do we tell whether the organisms we see under the microscope are dead or alive? This leads to another problem. When does a live organism become 'dead'? Fortunately, the question as applied to microbial life has no overriding philosophical or religious overtones and we can seek a useful answer. The only practical definition is to call an organism living (or viable) if it is capable of continuing division. It is dead (or non-viable) if it cannot multiply, although it may still be metabolically active. This brings us back to the original question—how can we tell if a cell is viable under a microscope? Staining methods have been developed that show a correlation with viability, but generally all that we can reasonably expect to do with a microscope is to obtain a total count of living and dead cells. In order to count only living cells we must measure directly the ability to multiply. This is the basis of a *viable count* which is normally done by dispersing a known volume of a cell suspension (diluted if necessary) onto a plate of a suitable growth medium and after incubation one colony will be produced from one viable organism. Viable counts done in this way are by far the most sensitive method of determining growth or multiplication since most others require at least a million cells to measure with any accuracy. Unfortunately, viable counts are tedious to do, and for many purposes may be unnecessary.

A variety of methods have been employed to measure growth as distinct from multiplication, for example the determination of the dry weight, the amount of a typical cell component, and the volume of packed cells or their catalytic power. The method actually adopted depends upon the purpose of the experiment. If rapidity is the primary consideration, by far the best is to measure turbidity by light scattering either directly in a nephelometer or indirectly with a spectrophotometer.

The batch culture of micro-organisms

Let us consider what happens when growth is measured after inoculation of a small number of suitable micro-organisms into a sterile liquid medium. A typical growth curve is given in Fig. 5.4; six main phases can be recognized.

1 *The lag phase.* This period, in which no growth or multiplication occurs, represents a time when the inoculum cells are adapting themselves to active growth in the new environment. The length of the phase varies widely but it will be long if the inoculum cells are old, are damaged in any way or if they have been grown previously in a quite different medium. If a spore inoculum has been used, the lag phase will also include the time required for germination. If the inoculum is of rapidly growing vegetative cells from an identical medium then the lag will be hardly perceptible.

2 *The acceleration phase.* Normally cells come out of lag asynchronously, giving rise to an acceleration phase where the population growth rate gradually increases as the proportion of cells growing increases. If we plotted biomass on our growth curve, the lag phase would appear shorter. This is

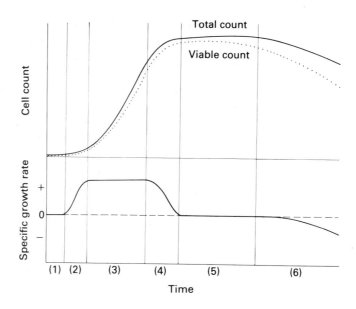

Fig. 5.4. The batch growth curve. The top section represents the cell count and the bottom section the specific growth rate. (1) lag phase; (2) acceleration phase; (3) exponential phase; (4) deceleration phase; (5) stationary phase and (6) death phase.

because cells must grow and double in size before they divide and multiply.

3 *The exponential phase.* Here the cells are growing at a constant rate, that is the generation time (the time between successive divisions) is constant. This is the only phase when the population is growing at a constant rate and it represents the maximum rate, and minimum generation time for the prevalent conditions. We therefore have autocatalytic growth, with the rate of increase in biomass (x) and cell number (n) proportional to the existing biomass or cell number. This may be represented by differential equations:

Rate of change of biomass

$$\frac{dx}{dt} = \mu x$$

Rate of change of cell number

$$\frac{dn}{dt} = \mu n$$

where t represents time and μ is the specific growth rate. We can integrate these equations to give the following expressions, which are of more practical use:

$$x = x_0 e^{\mu t} \qquad\qquad n = n_0 e^{\mu t}$$

$$\text{or } \log x = \log x_0 + \mu t \qquad \log n = \log n_0 + \mu t,$$

where x_0 and n_0 are values for x and n at the beginning of the exponential phase.

From these equations it follows that both biomass and cell number increase exponentially (see Fig. 5.3), so that although the actual increase per unit time in x or n becomes greater during this phase, the proportional increase is constant. Fig. 5.5 illustrates this and also shows how we can calculate experimentally the specific growth rate and generation time or doubling time of the population. The consequence of all this is that although the increase in cell number is slow initially, it eventually becomes explosive and the practical effects of such growth in nature are many. For example, once a bacterial infection is properly established in the body, any means of combating it must be increasingly efficient. The actual rate of exponential growth varies widely, some bacteria at high temperatures having a generation time as small as ten minutes while in some eukaryotic micro-organisms it may be as high as 24 hours. However, this exponential phase cannot go on indefinitely. Consider again the case of a typical bacterium with a volume of $1~\mu m^3$ growing with a generation time of 20 minutes. Fig. 5.5 shows that after 200 minutes the cell number has increased by a factor of 1 024. For the sake of convenience, let us round this figure to a thousand-fold increase. This means that as long as exponential growth continues, there will be a further

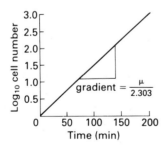

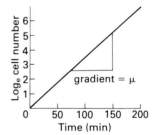

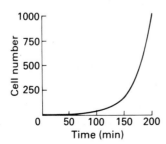

Fig. 5.5. The multiplication of a unicellular micro-organism with a doubling time of 20 minutes. A straight line is obtained with both semi-log plots but specific growth rate is calculated directly from the natural log ($\log_e$) plot. The doubling time $= \log_e 2/\mu = 0.693/\mu$.

thousand-fold increase every 200 minutes with the results presented below:

Time (min)	No. of cells	Volume of cells
0	1	$1 \ \mu m^3$
200	10^3	$10 \ \mu m^3$
400	10^6	$100 \ \mu m^3$
600	10^9	$1 \ mm^3$
1 200	10^{18}	$1 \ m^3$
1 800 (30 hours)	10^{27}	$1 \ km^3$
2 400 (40 hours)	10^{36}	$1 \ 000 \ km^3$
3 000 (50 hours)	10^{45}	$1 \ 000 \ 000 \ km^3$

4   *The deceleration phase.* After 50 hours incubation the volume of cells would be greater than that of the whole earth, thus illustrating the nature of exponential growth. Clearly, such results cannot occur in practice and growth must be circumscribed. The culture therefore enters the deceleration phase during which specific growth rate declines, usually as a result of the accumulation of toxic products of metabolism or changing environmental conditions, e.g. reduced pH. For example, many yeasts produce ethanol as a result of fermentation and as soon as a level of about 10% is reached, growth ceases and the cells may die. A practical consequence is that wines cannot be prepared with an alcohol content much above 14%; to increase this value a distillation stage must be introduced. During the deceleration phase, cells often start to produce secondary metabolites. These are compounds whose synthesis is not associated with growth and many are of industrial value, for example antibiotics.

5   *The stationary phase.* This phase is brought about through lack of food, or more accurately through exhaustion of one particular nutrient, or when accumulation of toxic products completely inhibits growth. In either case, biomass and cell number remain constant. Although the stationary phase can result from an equilibrium between multiplication and death, which may occur if toxic products accumulate, normally it is the consequence of neither growth nor death. In other words, the cells remain in a state of suspended animation. The length of the stationary phase depends upon the factor that initiates it and upon the micro-organisms concerned. It may last minutes, hours, days or even weeks. The biomass and cell numbers produced by the stationary phase depend to a large extent on the initial substrate concentration and on the organism's growth yield. This is the mass

of cells produced per unit of substrate converted and is often related to the efficiency of growth on a particular substrate. It is therefore of importance in industrial processes where the aim is to produce biomass as cheaply as possible.

6   *The death phase*. In this phase the cells begin to die and as a result there is an increasing divergence between the total and viable counts (Fig. 5.4). However, when death occurs, cells commonly start to digest themselves and lysis or, more strictly speaking, autolysis results. Such autolysis has some interesting results. If a laboratory culture is allowed to go into the death phase so that a reasonable percentage of cells lyse, those remaining may be able to grow on the products of this lysis. In other words, cannibalism occurs. This short growth phase is followed by a further death phase and the process is repeated until no cells remain. In this way a culture may remain viable for a very long time, although it should be noted that we are eventually selecting those organisms in a culture that lyse less readily or are the best cannibals. In evolution, micro-organisms have had to compromise between the requirement for maximal growth rates and an ability to adapt rapidly to the environment with a concomitant maximal ability to survive after the onset of the stationary phase. An excellent method for prolonging the stationary phase is to sporulate and so produce a cell specialized for the minimal rate of destruction in an environment inimicable to cell growth.

The continuous culture of micro-organisms

So far we have been considering the batch culture of micro-organisms. Growth occurs in a vessel of finite dimensions and eventually ceases. Is it not possible to devise an apparatus in which the cells are kept growing indefinitely by the continuous addition of fresh nutrient medium and the continuous removal of cells and their products? Surprisingly enough, work on these lines was not initiated until the 1940s and even then the significance of the pioneer work took some time to penetrate. Indeed, much of the development was the result of work on more efficient ways of producing cells for biological warfare (or, as we are usually told, for defence against biological warfare).

The method of growth was called continuous culture and the apparatus most commonly used was the *chemostat*, a simple form of which is illustrated in Fig. 5.6. Fresh sterile medium is pumped into the growth vessel at a steady rate while a constant level device removes cells and their products

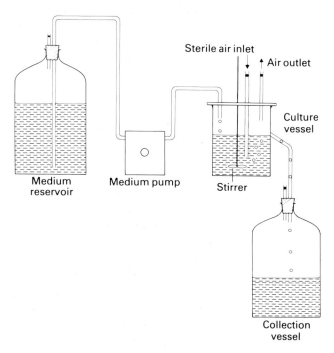

Sterile air inlet

Air outlet

Culture vessel

Medium reservoir

Medium pump

Stirrer

Collection vessel

Fig. 5.6. An apparatus for continuous cultivation of an aerobic micro-organism. Although not shown, the growth vessel normally contains means to maintain a constant temperature and pH and to prevent foaming.

at an equal rate. A stirrer homogenizes the fresh medium with the culture and normally also acts as an aeration device. The apparatus can be elaborated further by devices for sampling, pH control, oxygen control, temperature control, antifoam addition and so on. If we inoculate a chemostat and supply fresh medium at a constant rate, the rate of change in biomass concentration will depend on the increase resulting from growth and the decrease due to removal of cells in the effluent:

$$\frac{dx}{dt} = \underset{\text{growth}}{\mu x} - \underset{\text{removal.}}{Dx}$$

The symbol D represents dilution rate which is equal to the rate of flow of fresh medium divided by the volume of medium in the culture vessel. If D is less than the maximum specific growth rate of the organism, a *steady state* is eventually established, in which the actual specific growth rate equals the rate of removal, i.e. $\mu = D$. This occurs because specific growth rate becomes limited by the concentration of a single substrate and falls below its maximum value. The equation describing the relationship between substrate concentration (s) and specific growth rate (μ) is analogous to the Michaelis-Menten equation for enzyme kinetics. It has the form:

$$\mu = \frac{\mu_m{}^s}{K_s + s},$$

where μ_m is the maximum specific growth rate achieved when the substrate is in excess, and equivalent to the specific growth rate during exponential growth in batch culture. This equation is known as the Monod equation after the French microbiologist who first proposed it. The term K_s is the saturation constant for growth and is the substrate concentration at which μ is $1/2\mu_m$. The smaller the K_s value the greater the organism's growth rate at low substrate concentration. Bacteria usually have very low saturation constants, reflecting their ability to scavenge the low levels of nutrients often present in nature; this is one factor which determines competitive ability. The precise relationship described by the Monod equation is illustrated in Fig. 5.7 but in practice the curve will be slightly different. This is because as well as requiring energy for growth, cells must also maintain their structural integrity and carry out processes not directly associated with growth. The energy required for such processes is called *maintenance energy* and its inclusion means that the line in Fig. 5.7 will not pass through the origin, as even a non-growing cell will require nutrients to supply energy for maintenance.

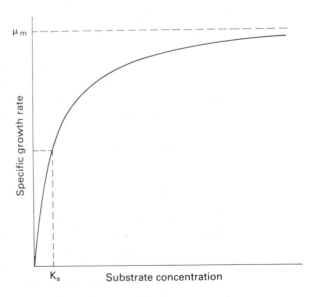

Fig. 5.7. The relationship between substrate concentration and specific growth rate as described by the Monod equation.

Chemostats therefore operate at sub-maximal specific growth rates where growth is substrate limited. In the steady state biomass concentration, substrate concentration, and everything else in the vessel remain constant, and in theory can do so indefinitely.

Continuous culture provides enormous advantages over batch culture, particularly in the research laboratory where reproducibility in the environment is more important than simplicity in technique, and also in industry where large amounts of cells or products are required at economic costs. Some of the advantages are as follows:

1 By altering the composition of the medium the substrate which is limiting growth may be changed and the specific roles of different nutrients can be investigated.

2 Environmental conditions and specific growth rate may be kept constant over extended periods and not just for the length of the exponential phase.

3 Growth at submaximal rates may be investigated and growth rate may be changed to a new constant value merely by altering the setting on the medium pump.

4 Effects of growth rate can be separated from those of environmental factors. For example, in batch culture growth rate can only be changed by altering factors such as temperature or medium composition which in themselves may change the properties and composition of a cell. In a chemostat, temperature can be increased or decreased but the actual specific growth rate will be constant as long as D is kept constant. Conversely, we can change D and μ while keeping temperature constant.

For these and other reasons, the chemostat has become a powerful tool in the study of microbial biochemistry, physiology and genetics and also in our understanding of the ecology of micro-organisms and their interactions. At the same time, it must be realized that the chemostat is a relatively sophisticated piece of apparatus if it is to be used properly and that contamination and/or evolutionary change over the long periods involved may be a problem.

Mycelial growth

The filamentous growth form results when cells do not separate after division and is most highly developed in filamentous actinomycetes and fungi. In both groups, extension of individual filaments, or hyphae, occurs by addition of material to the hyphal tips. In fungi, material is synthesized in a length of hypha called the *peripheral growth zone* and is transported to the tip in membrane-bound vesicles. Hyphae of the bread mould fungus, *Neurospora crassa*, extend relatively quickly, up to 38 μm/minute, and at this rate have a peripheral growth zone length of 4 mm. Approximately 38 000 vesicles must fuse with the tip each minute to maintain this extension rate. Material synthesized behind the peripheral growth zone will never catch up with the tip and must be accommodated elsewhere, so a branch is initiated and repetition of this process eventually leads to the development of a complex branching *mycelium*. Actinomycetes form very similar mycelia, providing an example of convergent evolution with similar growth forms evolving in two widely differing groups of organisms, one prokaryotic and the other eukaryotic.

Because cells do not separate in filamentous organisms, a cell cycle cannot be defined. Instead we have a duplication cycle in which the apical cell doubles its length, nuclear division takes place, septation separates the daughter nuclei and a branch is formed. In a sense, branch production is equivalent to cell division and mycelial organisms grow exponentially. Although individual hyphae extend at a linear rate, the number of branches increases exponentially, giving an exponential increase in biomass and total hyphal length. In liquid culture, filamentous organisms give a batch growth curve similar to that in Fig. 5.4, and also can be grown successfully in continuous culture. Deviations can occur, however, if mycelia form tight spherical bundles of hyphae, called pellets, or if differentiation and sporulation occur.

The filamentous growth form is adapted to colonization of solid substrates and surfaces. Unicells typically form small, self-limiting colonies on agar plates while a single fungal colony may cover the whole plate. This is because filamentous organisms can regulate the distribution of biomass between tip extension and branch formation. On nutrient-poor media, branch production is suppressed so that material may be directed towards hyphal tips to maintain rates of extension into regions of fresh nutrient. On nutrient-rich media, branch production is increased to enable more extensive colonization behind the growing margin. Unicells cannot regulate biomass production in this way and colonies are formed by cells pushing each other aside and piling on top of one other.

Synchronous growth

A growing microbial culture contains cells dividing asynchronously and the properties of the population are the average properties of the individual cells. When studying cell cycle events we want to measure changes in the biochemical properties of individual cells and we therefore need to amplify the physiological events by producing a synchronously dividing culture in which all the cells divide at roughly the same time. Two general methods of producing synchronous cultures are available.

1   *Induction methods*. These methods rely on synchronizing an exponential-phase culture by appropriate and usually sudden changes in the environment such as alteration in temperature, concentration of nutrients, or illumination for phototrophs.

2   *Selection methods*. Here there is a physical separation of cells from an exponential-phase culture at a particular point in the growth cycle. For example, centrifugation on a density gradient can separate the smallest cells from a culture and it can be assumed that these have all just divided. A common method involves filtration of cells through a cellulose nitrate filter which is then inverted, so that the cells are on the lower surface, and liquid medium is passed through the filter from above. When cells divide they fall off the filter giving a continuous supply of newly born cells. Other methods depend on filtration of cells of certain sizes or the selective adsorption of cells on surfaces. Although producing fewer cells, these selection methods are preferable to those involving induction as they are less subject to the objection that the process of synchronization causes an extensive metabolic disturbance within the cells and hence produces abnormal results.

Fed-batch culture

Although batch and continuous culture are used most frequently in the laboratory and are most useful for experimental purposes, industrial processes often use a hybrid of the two techniques called fed-batch culture. This involves ordinary batch growth until the substrate is fully utilized, followed by regular additions of fresh substrate which is then rapidly metabolized. The total culture volume may be allowed to increase or portions may be removed as fresh substrate is added. This type of culture was developed empirically but microbiologists are now beginning to understand the physiology of growth under such conditions and the reasons for its efficiency in terms of product formation, yield and greater control. This increased understanding will hopefully lead to further advances in productivity.

6 : Viruses

Viruses are a group of microscopic infectious agents which are distinct from the two basic forms of cellular life. Since they are non-cellular and have no associated metabolism one can argue that viruses are not organisms at all, but rather forms of a complex molecule which enslave the cells they infect in order to replicate themselves. Since all viruses are obligate parasites of cellular organisms they almost certainly evolved after cellular life had developed, and therefore do not represent a primitive, ancient form of prokaryotic or eukaryotic life.

The structure of viruses

A few of the largest viruses are bigger than the smallest bacteria, although most are very much smaller and cannot be resolved using a light microscope. Consequently, their existence was first inferred by the ability of extracts of diseased tissues to retain their infectivity even after they had been passed through a filter with a porosity low enough to remove cellular organisms (c. 0.2 μm pores). In other words, the term 'virus' was synonymous with a filter-passing agent.

It was shown that viruses attacked a wide range of living organisms in the microbial, animal and plant kingdoms. Growth could be measured in terms of an increase in infectivity and it was soon found that it only occurred within a host. Methods were developed to purify the virus particles and by 1935 Stanley had obtained a sample of the tobacco mosaic virus in a crystalline state. This was a revolutionary step, showing that many of the properties of viruses were similar to those of high molecular weight polymers studied by biochemists; similar methods could therefore be used in their

study and purification. With the advent of the electron microscope it became possible to study the morphology of viruses
and to correlate the results with chemical analysis of purified
preparations.

The usual method of naming viruses is to describe the host
and often the symptoms of the disease caused. A virus
attacking tobacco plants and causing bleaching of
chlorophyll in spots on the leaves is thus called tobacco
mosaic virus; one isolated from the adenoids of a man is
called a human adenovirus; one attacking *Escherichia coli* is
called an *Escherichia coli* bacteriophage, or phage for short.
Different strains may be given letters and numbers such as the
series of T-phages (Fig. 6.3 a,b) associated with many basic
experiments in molecular biology. The reason for this classification is that an alternative system based on morphology
would be almost valueless. Indeed, viruses attacking quite
different hosts may appear morphologically indistinguish-
able, for example the human poliomyelitis and the turnip
yellow mosaic viruses are very similar when considered
simply on morphological characteristics.

The size and shape of viruses

As soon as it became possible to view viruses in the electron
microscope, three generalizations emerged.
1   A wide variety of different shapes and sizes occurred,
ranging from the f2 phage with a roughly spherical shape and
a diameter of 20 nm, through the complex tailed phages to the
vaccinia virus with a brick shape of 250×300 nm.
2   The individual particles of a particular virus were
identical in size and shape. The revolutionary implications of
this fact were only slowly realized by most virologists; the
startling conclusion being that viruses do not grow and
multiply but must somehow be formed *de novo*.
3   The use of negative-staining methods led to an overall
picture of virus structure. Each virus particle (virion) consists
of a core of a single species of nucleic acid enclosed by a
proteinaceous coat or *capsid* which is itself made up of individual protein subunits called capsomeres. The capsid is in
some cases surrounded by an *envelope* of lipid and protein.
On the basis of the way in which capsomeres are assembled to
make up a capsid and of the presence or absence of an
envelope, it is possible to distinguish four main types of virus
organization as illustrated in Fig. 6.1, as well as a few more
complex designs.
Naked helices. In the helical viruses, the individual capsomeres are arranged spirally so that the final capsid is a

89

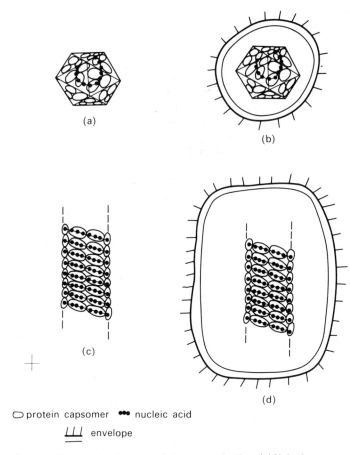

Fig. 6.1. The four basic types of virus organization. (a) Naked icosahedral, (b) enveloped icosahedral, (c) naked helical and (d) enveloped helical.

hollow tube. The individual capsomeres are identical and the nucleic acid is located in a helical groove on the inside of the cylindrical capsid. An example is shown in Fig. 6.3e,f.

Enveloped viruses. Either the helical or the icosahedral type of virus is sometimes surrounded by a loose membranous envelope. Most enveloped viruses attack animals and the envelope seems to be partly derived from the host plasma membrane. Examples are shown in Fig. 6.2b,c.

Complex viruses. A few have a more complex structure than those described above. The best analysed examples are the tailed bacteriophages and, in particular, the coliphages (Fig. 6.2a and Fig. 6.3a,b). Whereas the head of the phage is an almost conventional icosahedron containing the viral nucleic acid, there is a very complex tail structure attached to it containing at least five different protein types making up the sheath, core collar, end plates and fibres. However, all these

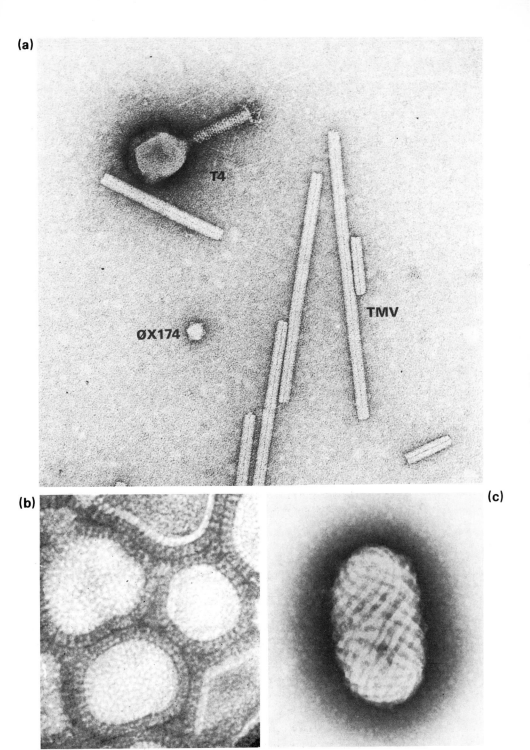

Fig. 6.2. (a) Negatively stained micrograph of Tobacco Mosaic Virus, φX174, and bacteriophage T4 (×120 000), (b) enveloped human influenza virus (×320 000), and (c) Orf virus (×240 000). With kind permission of: (a) E. Eiserling; (b) R.W. Horne and (c) D. Gregory.

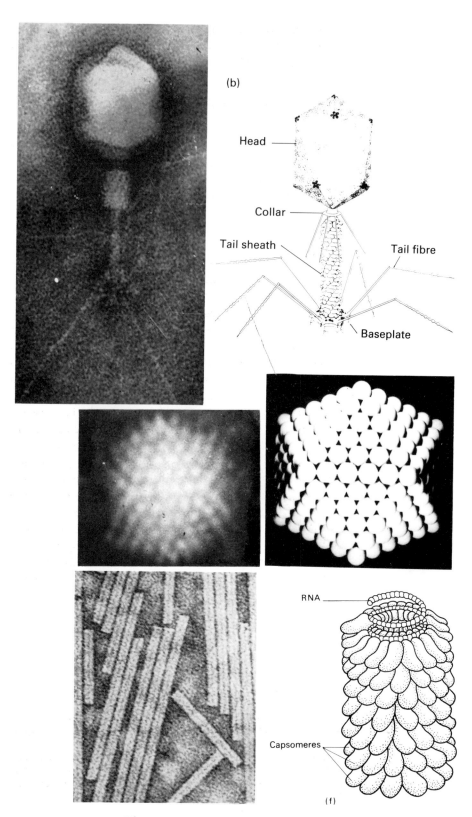

(b)

Head

Collar

Tail sheath

Tail fibre

Baseplate

RNA

Capsomeres

(f)

structures are formed by the condensation of sub-units by a process analogous to crystallization. The importance of this will be realized when virus replication is discussed later.

The chemical composition of viruses

There are two basic chemical components in viruses—protein and nucleic acid. In many viruses they are the sole components. It is only the nucleic acid which is necessary for viral production, the protein being concerned in the transference of the nucleic acid from one cell to another.

Viral proteins contain the same amino acids in similar proportions to those of cellular forms of life. Most of the protein has a structural function and makes up the individual capsomeres which are themselves usually made of identical or nearly identical subunits.

Naked icosahedrons. Careful study of the so-called spherical viruses showed that their capsids are all icosahedral in shape, that is they are made up of a structure with twenty triangular faces and twelve vertices. The icosahedron is made up of protein capsomeres which are arranged according to the laws of geometry and crystallography. Vertex capsomeres are built from five protein subunits and the capsomeres of faces from six proteins. All icosahedral viruses have the basic twelve vertices while larger viruses have more face-building subunits inserted in between the vertices. Thus, a range of sites of icosahedral viruses is possible using the same geometrical shape (See Fig. 6.2 for examples of two icosahedral viruses of different size). You can cut out and build an icosahedron from the pattern provided at the back of this book. It is fascinating to note that exactly similar principles are used in the building of geodesic domes where a very strong design can be obtained from quite light and easily assembled units.

The nucleic acid of icosahedral viruses is located in a very condensed form inside the capsid. An example of a naked icosahedral virus is shown in Fig. 6.3c and d.

Fig. 6.3. (a) A T2 coliphage (×320 000) and (b) diagram of the structure of bacteriophage T4 (length 200 nm). (c) Negative stain of adenovirus (×560 000) and (d) model showing the same virus icosahedral assembly. (e) Helical tobacco mosaic virus (×144 000) and (f) diagram of TMV showing helical arrangement of protein capsomeres and RNA genome. (a) From S. Breener *et al.*, 1959; (b) with kind permission of E. Eiserling, (c, d and e) R.W. Horne and (f) D.L.D. Caspar.

The reproduction of viruses

The system that has been generally used to study the reproduction of viruses is the attack of a bacterium by a bacteriophage. Indeed, much of the work has been confined to a series of phages attacking *Escherichia coli* called coliphages. They were chosen arbitrarily and called by the letter T and a number. In practice, three 'T-even' phages (T2, T4 and T6) having similar properties have been the ones chosen by most investigators. Although they have a rather complex structure for a virus (see Fig. 6.3b), they have proved to be very suitable for experimental purposes and have provided us with a wealth of information, most of which can be applied to viruses in general.

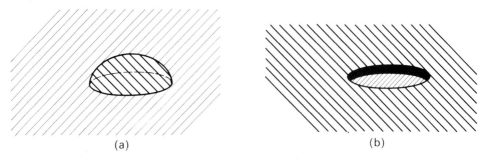

(a)   (b)

Fig. 6.4. A comparison of a bacterial colony with a phage plaque; (a) a bacterial colony growing on an agar plate, and (b) a phage plaque on a confluent agar lawn.

The counting of phages

It is difficult and unreliable to count viruses in the electron microscope and instead a viable count is normally carried out, the measure of viability being the ability of a phage to infect a sensitive bacterium, to multiply inside it, and eventually to lyse it, liberating fresh phages which attack neighbouring cells. In practice, a suspension of sensitive bacteria and phages are first mixed together to allow the phages to absorb into the bacteria. The mixture is then diluted with a soft agar and spread over the surface of a plate. Non-infected bacteria will grow and form a background lawn of growth on the agar.

Where single phages have absorbed onto cells they will infect, multiply, lyse the bacterium and eventually form a clear area or plaque on the plate where all the bacteria have been killed. This plaque is a virus colony (Fig. 6.4) and since it is produced from infection of cells with single, viable virions, it can be used in the same way as colonies of cellular micro-organisms for counting purposes. The number of infective virus particles in a suspension can therefore be expressed as the number of plaque forming units per millilitre (PFU/ml).

The one-step growth curve

The essential features of virus multiplication can be seen in the one-step curve experiment. A culture of sensitive bacteria is mixed with a relatively small number of phage particles and incubated for a short period to allow adsorption of the phages onto the bacteria. The culture is diluted, allowed to grow and the number of phage particles plus infected bacteria is measured. It should be noted that a single plaque will be formed from either a phage particle or an infected bacterium irrespective of how many phages the latter may contain within it. Typical results are given in Fig. 6.5.

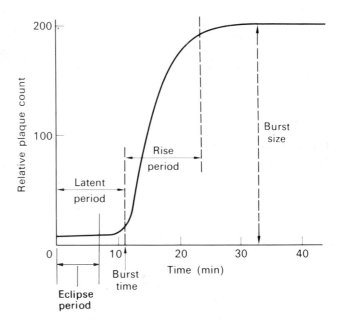

Fig. 6.5. A one-step growth curve for a phage attacking a bacterium.

Initially, there is a *latent period* in which there is no increase in the number of plaques. Beginning at the same time as the latent period, but spanning a shorter duration, is the *eclipse period*. This encompasses the time taken to synthesize and assemble new virus particles and corresponds to a period during which one cannot recover any complete viruses from the infected cells. At the end of the latent period there is a rapid rise in plaque-forming units as the bacteria begin to lyse, liberating new phages. This *rise period* eventually levels

95

out, the plaque count remaining more or less constant because the initial dilution of the culture largely prevents adsorption of newly liberated phages on the remaining uninfected bacteria. The factor by which the number of phage particles increases the experiments is known as the *burst size* and it represents the average number of phages produced on the lysis of one infected bacterial cell. The potential of virus multiplication is evident since the burst size is usually over a hundred and the whole process represented in the one-step growth curve may take only 10–20 minutes.

The multiplication of a phage

The multiplication of a phage, as typified by the T-even coliphages, can be considered in three phases:
1   Adsorption and penetration of the host.
2   Synthesis and assembly of new phage within the host.
3   Lysis and liberation from the host.
1   *Adsorption and penetration of the host cell.* If a suspension of virus particles is mixed with sensitive bacteria, the phages attach themselves to the cell surfaces (Fig. 6.6a). This adsorption is a very specific process and can be likened to that between an enzyme and its substrate or to an antibody and the corresponding antigen. There are complementary sites in the phage tail fibres (see p. 92) and in the bacterial cell wall. The specific adsorption is followed by a further adsorption of the tail fibres onto the surface of the cell wall (Fig. 6.6b). This is followed by a contraction of the tail sheath and the forcing of the tail core through the cell wall and cytoplasmic membrane (Fig. 6.6c). Finally, the phage DNA enters the host cell (Fig. 6.6d) by a process analogous to injection by a hypodermic syringe, although there is no actual contraction of the phage head; it is not known what causes the DNA to pass from the phage to the bacterium.

The end result of the above mechanism is that the protein capsid is left behind on the outside of the host cell while the nucleic acid passes in. This fascinating result was fore-

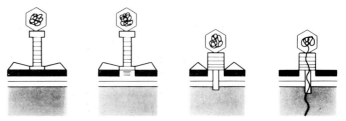

Fig. 6.6. The adsorption of a phage onto a sensitive host bacterium and injection of the phage nucleic acid (for full description see text).

shadowed by the finding that if phage protein is specifically labelled with radioactive ^{35}S and phage DNA with radioactive ^{32}P, then only the ^{32}P passes into the bacterial cytoplasm and plays any part in the production of further virus particles. This proved to be a key experiment in the development of molecular biology since it showed that only the phage nucleic acid was necessary to code for the production of complete mature phages. In other words, all the genetic information was contained in the phage DNA. Another striking illustration of the fact that the protein capsid is solely concerned with the passage of the infectious nucleic acid from one cell to another, is the observation that naked phage DNA itself can be used to transform cells (See Chapter 7) and lead to the production of normal phages.

2   *Synthesis and assembly of new phages.* The entry of the phage DNA into the host cell causes an almost immediate reduction in the rate of the synthesis of host nucleic acid and protein. The metabolic machinery of the bacterium is turned over to the synthesis of polymers coded for by the phage DNA, in which three phases can be distinguished.

(a)   Under the influence of host RNA polymerase, a part of the virus DNA is transcribed to mRNA. This 'early' mRNA is then translated by host ribosomes to produce 'early' proteins. These proteins are enzymes: although viral reproduction depends in the main upon host enzymes already present in the cell at the time of infection, additional virus-coded enzymes are essential or stimulatory to the later replication of phage DNA. For example, the T-even phages code for the synthesis of as many as twenty of such early enzymes. However, it must be emphasized that these proteins are not structural components and are not incorporated into the mature phage particle.

(b)   The second phase consists of the synthesis of the structural protein and DNA of the phage. We have seen that the T-even phages have a relatively complex structure and that there will be many different protein subunits required. Such 'late' proteins are synthesized from 'late' mRNA produced by regions of the phage genome not transcribed in the previous phase. At about the same time, new viral DNA is produced. We have thus produced all the components necessary to form the mature phage particle.

(c)   The final stage is the assembly of these components. This assembly does not occur randomly with respect to time but instead occurs in a definite sequence or morphogenetic pattern. The viral DNA molecules condense under the influence of a 'late' protein called the condensing principle, while the capsid subunits undergo *self-assembly* by a process

97

analogous to crystallization to form empty heads. The condensed DNA then passes into the empty heads. Similarly, inevitable processes lead to phage tail and tail fibre assembly; all three then combine with the heads to give intact phage particles.

The precise order and pattern of assembly of phage head and fibres has been worked out using strains of phage which are temperature-sensitive mutants (see Chapter 7). These mutants produce non-functional, incomplete phage particles when cultured at a high (*restrictive*) temperature and undergo normal lytic development at a lower (*permissive*) temperature. If bacterial cells are infected at the restrictive temperature and then lysed, the precise morphology of the mutant can be determined in the electron microscope. Different mutants have been found to be defective at different steps of the assembly pathway producing a range of partially completed virus particles. In this way a sequence of morphogenetic events has been mapped out in which phage heads, tails, and tail fibres are made separately; tails are then added to heads before the tail fibres are added to the head-tail assembly (Fig. 6.7). Further experiments showed that incomplete phages of certain pairs of mutants could be mixed together in a test tube to yield fully formed complete phage particles. The ability of a pair of mutant strains to exhibit *complementation in vitro* meant that these strains were blocked in different steps of the assembly pathway and that the mutation in one strain was being made good by the presence of a wild type component in the other. The number of steps in the assembly pathway could thereby be determined from the number of mutant types that were able to exhibit *complementation in vitro*. Not all of the genes that are involved in the morphogenesis of a bacteriophage such as T4 code for the virus particle. Other genes appear to be necessary for the proper sequencing of the assembly process and in packaging of the DNA into the capsid.

The production of new phages can be summarized as follows:

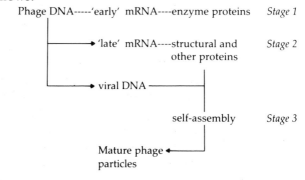

Phage DNA-----'early' mRNA----enzyme proteins   *Stage 1*

'late' mRNA----structural and   *Stage 2*
other proteins

viral DNA

self-assembly   *Stage 3*

Mature phage particles

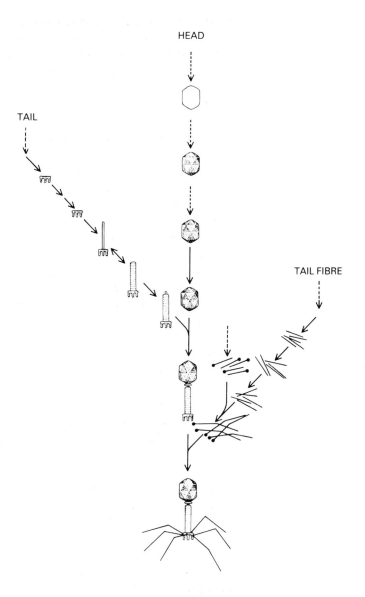

HEAD

TAIL

TAIL FIBRE

Fig. 6.7. Pathway for self-assembly of head capsid, tail and tail fibres of bacteriophage T4 (for full description see text). With kind permission of W. Wood.

3 *Lysis and liberation from host.* Associated with the assembly of mature phage particles is the production in the host cytoplasm of another late viral protein with a lysozyme-like activity similar to that found in the phage tail. This enzyme catalyses the partial hydrolysis of the bacterial peptidoglycan causing a weakening of the cell wall and an eventful osmotic rupture. The phages are thus liberated into

the environment where they can infect other cells and start the whole process again.

Although this description of the phases of virus multiplication is for the more complex T-even coliphages, in general it applies to all viruses. However, there are variations in the ways in which viral genomes are replicated (see below) and in the way in which the nucleic acid enters the cell. For example, in the simple plant viruses there is no specific adsorption or entry mechanism but passage occurs through breaks or cuts in the plant surface and is frequently insect-mediated. Animal viruses are carried into the cells by phagocytic or pinocytic action and it is only then that the nucleic acid becomes separated from the capsid; release often occurs without lysis by a process analogous to a reversed pinocytosis.

Replication and expression of viral genomes

Viral genomes can be of RNA or DNA, which can be double or single stranded and circular or linear. These may code for as few as three genes in the case of the RNA virus R17, to more than a hundred genes in some of the complex T-even bacteriophages. In some small viruses such as the bacteriophage ϕX 174 the information content of the DNA is increased by having the genes overlapping on the DNA.

Given the diversity of types of viral genomes it is not surprising that the pattern of genome replication and gene expression differs markedly in different groups. In some cases the RNA or DNA strands can act directly as templates for translation or transcription while in others the complementary template must first be synthesized before the information encoded in the genes can be read. For example, single stranded RNA viruses such as poliovirus have an RNA which acts directly as an mRNA for the synthesis of protein and also as a template for the production of a complementary RNA strand from which new viral genomes can be transcribed (Fig. 6.8a). Other RNA viruses such as the retroviruses produce an intermediate DNA strand from their RNA genomes by an enzyme called *reverse transcriptase*. This DNA then becomes integrated into the chromosomal DNA of the host cell where it acts as a template for the synthesis of viral mRNA and genomic RNA (Fig. 6.8b). Integration of a retrovirus genome into a mammalian cell chromosome may lead to a recipient cell undergoing uncontrolled rapid division; that is, it may become cancerous or *transformed*. The agent responsible for AIDS (Acquired Immune Deficiency Syndrome) may be a retrovirus that is closely related to human T-cell leukaemia

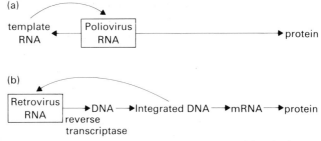

Fig. 6.8. Replication of and protein synthesis from (a) poliovirus and (b) retrovirus RNA genomes.

virus. The latter is known to cause lymphomas by transformation of white blood cells; cancers associated with AIDS may therefore have a similar aetiology. A variety of double stranded DNA viruses of animals, including adeno papova and herpes viruses, have also been shown to transform mammalian cells after integrating into the host genome. Although we cannot yet clearly define the role of viruses in causing cancers, we know that the sequence of nucleotides in the genomes of many cancer causing (or oncogenic) viruses is very similar to the sequences found in human 'oncogenes', which are responsible for the cancerous transformations of human tissues which are not infected by viruses. Some bacteriophages are also able to incorporate their genomes into the chromosomes of their bacterial hosts. This process is known as lysogeny.

Lysogeny

In *virulent* phages such as bacteriophage T4, the entry of the viral nucleic acid leads to an irreversible series of events culminating in cell lysis and the liberation of the progeny. However, an alternative mechanism exists in many viruses by which the DNA, after entry into the host cytoplasm, becomes integrated into the host genome and reproduces in phase with it in a form known as a *prophage* (see Fig. 6.9). This virus–bacterium relationship is called lysogeny and those phages that can be integrated are called *temperate* (to distinguish them from the virulent ones discussed above). A few phages such as P_1 may exist in the bacterial cell either as a non-integrated independent plasmid or, at higher temperatures, as an integrated lysogen.

We know the prophage continues to be reproduced in synchrony with the lysogenized bacterium for two reasons:
1   Occasionally there is a spontaneous initiation of a lytic cycle of development, the rate of which can be increased by certain inducing agents.

101

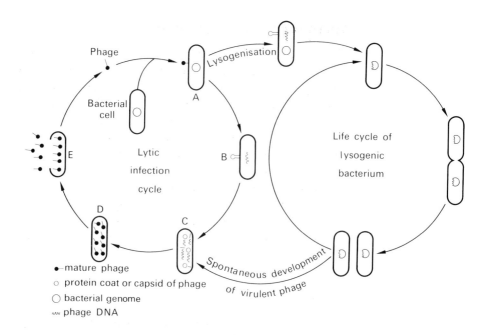

Phage

Bacterial cell

Lysogenisation

A

Lytic infection cycle

B

E

D

C

Life cycle of lysogenic bacterium

Spontaneous development of virulent phage

•—mature phage
○ protein coat or capsid of phage
◯ bacterial genome
∿∿ phage DNA

Fig. 6.9. The cycles of lytic infection by a virulent phage and lysogeny by a temperate phage.

2 The phenotypic properties of a bacterium are often changed by the presence of prophage. In particular, the lysogenized bacteria become immune to further attack by related phages. However, other properties of the host bacterium may also be altered during lysogeny. For example, non-toxigenic strains of the diphtheria bacillus (*Corynebacterium diphtheriae*) are transformed to virulent toxigenic variants after infection by phage B. Variations in the chemical structure of the surface antigens of *Salmonella* can also be associated with lysogeny of certain phages. This phenomenon is referred to as *lysogenic conversion*.

When prophage DNA becomes excised from the bacterial genome, for instance during the induction of lytic development, mistakes can be made so that a small piece of host DNA may remain attached to the phage DNA. The host DNA can then become incorporated into the assembled virus and transferred with the viral genome into a new host. In fact, both virulent and temperate bacteriophages have been useful agents for the transfer of DNA from one bacterium to another. This process is called *transduction* and is widely used in genetic studies of bacteria (see Chapter 7).

Micro-organisms and viruses

The majority of prokaryotes can be attacked by a suitable virus. Some bacteriophages have the complex structure of the T-even coliphages, others have a simpler naked icosahedron shape, while a few are filamentous. Viruses also attack blue-green bacteria and show similar characteristics to bacteriophages. The position regarding viral infection of eukaryotic micro-organisms is less clear and the only well-established instances are some viruses attacking fungi. This may have commercial or economic ramifications, for example, in the mushroom industry. However, there can be little doubt that viruses do not commonly attack yeasts since such infections would almost certainly have been observed as a result of the widespread use of yeast in industry.

Viroids

These are minute infectious agents of plants which are composed of single naked RNA molecules. The first reports of these pathogenic RNA molecules in the early 1970s aroused considerable interest and a fair amount of scepticism since it was previously thought that infectious nucleic acids could only survive outside a cell if they were encapsulated in a protective protein capsid. Viroid RNA molecules are unusal in a number of respects. They are circular, uniquely folded and so small that the largest one so far described (CEV: citrus exocortis viroid) is only 371 nucleotides long—about one tenth of the size of the smallest RNA virus. Although viroids are now known to cause disease in over a dozen plants, they have not so far been implicated in any diseases of animals.

Prions

A group of recently discovered infectious particles which have attracted equal or greater controversy than viroids are the slow viruses or prions (proteinaceous infective particles). Prions are associated with certain degenerative diseases of the central nervous sytstem such as kuru, a disease restricted to a few cannibalistic tribes of New Guinea and transmitted by eating uncooked brain, and other disease such as scrapie of sheep and goats and Creutzfeldt-Jacob disease of humans and animals. Prions would seem to be proteins or glycoproteins with no associated nucleic acid of any kind. Since replication of the prion-protein presumably occurs by the normal means, this implies that these proteins may be able to bring about the production of their own requisite mRNA. If this were found to be so, prions would have a revolutionary impact on molecular biology since they would constitute the only example to date where genetic information passes from protein to nucleic acid and not the other way around. However, recent evidence indicates that scapie glyco-protein may in fact be a product of a normal host gene, indicating that these infectious agents may be more conventional than first thought.

7 : The Genetics of Micro-organisms

Our understanding of the mechanisms by which characteristics of micro-organisms are determined and are stably
maintained has increased dramatically over the last twenty-
five years. The remarkable turnabout in our understanding of
the structure and organization of the genetic material (DNA)
of cells has come about largely through the study of the
genetics of bacteria and also to a significant extent the
genetics of fungi. Indeed, microbes have become the organisms of choice in much work on fundamental problems in
molecular biology. As a result we now understand the functioning of the genome at the molecular level and are
increasingly able to manipulate cells to obtain improvements
in specific properties. This last ability is one of the corner-
stones of the new phase of industrial exploitation of bacteria
and fungi called biotechnology (see Chapter 10).

In the last thirty years it has come to be recognized that
there exists within biology an immutable relationship which
is responsible for the form of all organisms. Thus, the heri-
table characteristics are coded on the molecule, DNA, in
discrete units known as *genes*. During reproduction of one
cell to give two daughter cells the DNA is very precisely
duplicated such that each cell has an identical set of genes.
This process of *DNA replication* is essential for the maintenance of the identity of the daughter cells. Decoding of the
genes is carried out in two steps called *transcription* and
translation giving rise to the polymers RNA and protein,
respectively. Both processes are carefully controlled to ensure
the accurate transmission of information encoded in the
genes. Thus, information transfer almost always occurs
DNA > RNA > protein. This relationship is depicted in Fig.
7.1.

Certain bacteriophages and viruses which carry their genes in the form of RNA can direct the synthesis of DNA from RNA, and this is indicated by the dotted line in Fig. 7.1.

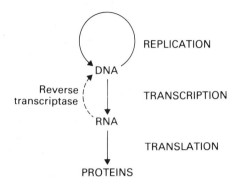

Fig. 7.1. Information transfer in cells.

The discovery that DNA is the hereditary material was made with the bacterium *Pneumococcus*, which when isolated from a patient with pneumonia produces a sticky mucoid capsule from the synthesis and secretion of a hygroscopic capsular polysaccharide. Subculture in the laboratory commonly results in the selection of non-mucoid colonial *mutants* which no longer produce capsular polysaccharide. These mutants have lost the capacity to cause pneumonia, they are thus said to be *avirulent*. It was shown that avirulent bacteria acquired virulence when incubated with DNA from virulent strains. No other macromolecule from the virulent cells possessed this capacity to *transform* cells and thus the link between DNA and inheritance of characteristics was established.

What is a gene?

The unravelling of the relationship between DNA and the characteristics of an organism derives from studies of both fungi and bacteria. The key proposal that genes act by controlling the specificity of protein synthesis arose from work with mutants of the fungus *Neurospora crassa* and led to the proposal that

1 gene = 1 enzyme.

Bacteria and, in particular *Escherichia coli*, have been used to define this relationship more precisely. Thus:

1 gene (or cistron) = 1 polypeptide.

106

This definition reflects the discovery that not all polypeptides are enzymes. Some play a structural role (e.g. flagellin, the main protein of the bacterial flagellum), and others are regulatory (e.g. the catabolite repressor protein, CRP; see below). The decoding of the information in a gene to give a polypeptide involves two processes, transcription and translation. Transcription is the process of copying the DNA to give an intermediary molecule, called the messenger RNA (mRNA), and is carried out by the enzyme RNA polymerase. During translation the RNA is translated into a sequence of amino acids which constitute the polypeptide. Translation requires the presence of two other types of RNA molecule, namely ribosomal RNA (rRNA) and transfer RNA (tRNA). Each of these three RNA molecules performs a different role. Translation takes place on the ribosome which consists of approximately 50 proteins bound together by rRNA molecules. The rRNA also binds the mRNA to the ribosome to initiate translation. Decoding of the message is essentially the role of the tRNA molecules which carry the amino acids to the ribosome and ensure that the sequence of their polymerization to give the protein is exactly that prescribed by the gene (Fig. 7.2).

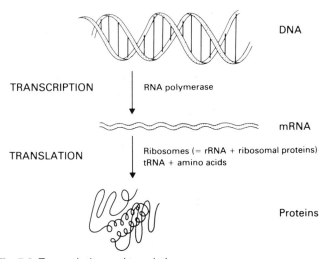

Fig. 7.2. Transcription and translation.

These processes are universal and are essentially identical in all organisms.

Control of transcription

The majority of genes, then, produce a specific polypeptide which may act directly or in combination with other peptides

107

to form enzymes or structural proteins. Such genes are called *structural* genes. We have also mentioned that bacteria are capable of controlling the synthesis of specific proteins by such processes as induction and repression (see pp 60, 71–72). A study of these processes has led to the view that in bacteria the structural genes controlling the synthesis of physiologically related proteins usually occur as a cluster on the chromosome, the region being called an operon. Such multigenic operons produce single molecules of mRNA initiated by the binding of RNA polymerase to a promoter sequence at one end of the operon. Situated immediately adjacent to the promoter sequence and sometimes over-lapping with it are the regulatory sequences which control the initiation of transcription. These sequences are the site of action of polypeptides produced by *regulator* genes (see Fig. 7.3).

A. Positive regulation

operator   promoter   structural gene

B. Negative regulation

Fig. 7.3. Generalized structure of a gene. Note that in negative regulation the promoter and operator may overlap.

promoter   operator   structural gene

There are basically two types of control of transcription, positive and negative.

Negative control

The basic action of these regulatory polypeptides is to combine within or close to the promoter at a region known as the operator. Binding of the regulator protein to the operator prevents the binding of RNA polymerase to the promoter. Synthesis of RNA is thus inhibited and no synthesis of the proteins coded for by the operon occurs. In other words, control is essentially by competition between the RNA poly-merase and the regulatory protein for the promoter (Fig. 7.4).

Positive control

Positive control has been recognized as a major mechanism of regulation only in the last ten years. In this type of regulation the initiation of transcription is *dependent* upon the binding of the regulator to the promoter. There are now many examples of this type of regulation but two examples serve to differentiate two classes of positive regulation. The catabolite

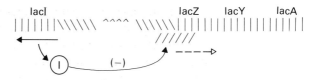

Fig. 7.4. The structure of the lactose operon of *E. coli.* The types of hatching correspond to the promoter (\\\\\), the operator (/////) and the structural genes (|||||): the binding site of the catabolite regulator protein is shown (vvv). The arrow indicates the direction of transcription of the genes. Note that in this case the operator overlaps the promoter and the start of the first structural gene. (Diagram not to scale).

repressor protein, CRP, referred to earlier, forms an active complex with the small molecule, cyclic AMP (cAMP), which acts on many different operons to increase the rate of their transcription (over 200 genes are regulated by the CRP–cAMP complex in *E. coli*). This is termed a 'global' regulatory system since it affects genes of many different metabolic systems. It is now known that there are many such global regulatory networks in organisms like *E. coli.*

The second type of positive regulator is one which is specific to only one operon or group of operons. When a number of operons are under the control of a single specific regulator it is termed a *regulon*. An example of this, the arabinose regulon *E. coli,* is shown in Fig. 7.5. Three operons specifying genes for the transport and metabolism of arabinose are regulated by the product of the *araC* gene. Binding of arabinose to the regulatory protein leads to the formation of a complex which can bind to the regulatory region of the regulon and stabilize the binding of RNA polymerase to allow the initiation of transcription.

Regulation of gene expression usually requires the presence of another molecule other than the protein produced by the regulator gene. This is usually either the *inducer* or the *repressor* depending on its mode of action, and the processes

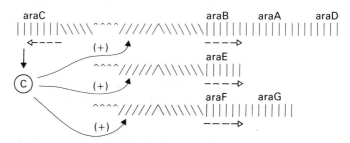

Fig. 7.5. The structure of the arabinose regulon of *E. coli.* The hatching pattern is the same as for Fig. 7.4. (Diagram not to scale).

109

of regulation are known as induction or repression, respectively. Both can be found in either positive or negative regulatory systems (Fig. 7.6).

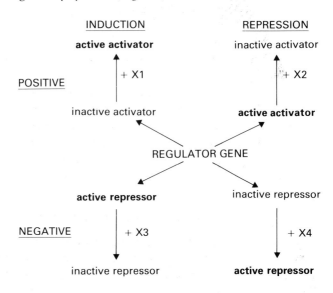

Fig. 7.6. Induction and repression in the context of positive and negative regulation. Examples of the regulatory molecules from *E. coli* are: X1, arabinose; X2, cysteine; X3, allolactose (the inducer of the lactose operon); and X4, tryptophan.

Induction

The inducer specifically combines with the regulatory protein rendering it either effective at binding to the regulatory sequence and thus stimulating transcription (positive regulation) or incapable of acting on the operator sequence and thereby switching on induced protein synthesis (negative regulation).

Repression

Binding of the repressor to the regulatory molecule either prevents interaction of the positive regulator with the regulatory sequence or increases the affinity of the negative regulator for the operator sequence (negative regulation) and thus stops transcription (Fig 7.6).

One of the most notable features of micro-organisms is their capacity for rapid adaptation to new environments. The speed of response to environmental stimuli depends upon the rapidity of synthesis of new mRNA molecules and the control of translation of existing mRNA molecules. In bacteria such molecules have an average lifetime of as little as 2−3 minutes and this allows for rapid shifts in the direction of protein synthesis. The mechanisms described above are but two examples of how expression of genes can be controlled; they provide a model for the type of system that may operate in higher cells and organisms. The last ten years have revealed many more types of regulation. These mechanisms are involved in processes as simple as amino acid biosynthesis to more complex processes such as switching on and off the synthesis of specific proteins in sporulation (p. 27).

Studies like these on the molecular basis of gene expression are now well known but they have only been made possible by the availability of suitable mutants and methods for recombination in prokaryotes. The rest of this chapter will be concerned with mutagenesis and recombination.

Mutation in micro-organisms

Mutation is a naturally occurring phenomenon leading to a large measure of variability within any population of cells. In organisms with only a single copy of each gene (*haploid* organisms) the occurrence of a mutation in a gene will lead to a change in the properties of the organisms. Mutation is a change in the DNA sequence of a gene and is said to lead to a change in the *genotype* of the organism. The ability of the experimentalist to see that change depends upon there being a change in the *phenotype* of the organism. Some typical examples of phenotypes are given below:

1 *Auxotrophic mutants.* We have seen that many micro-organisms such as *Escherichia coli* can grow on a medium containing a single carbon and energy source. They are called *prototrophs*. If a mutation occurs which results in the loss of the ability to synthesize an essential metabolite such as an amino acid or a growth factor, it will express itself in a nutritional requirement for the substance. Such mutants are called *auxotrophs* and were essential to the elucidation of the relationship between genes and enzymes. They have also been useful as markers for genetic experiments in the construction of genetic maps of different organisms.

2 *Resistant mutants.* Resistance to antibiotics and to phage occurs spontaneously and through a range of mechanisms, e.g. loss of cell surface components which act as phage receptors, and acquisition of enzymes able to metabolize the antibiotic. Acquired resistance of this type can have serious consequences since it means that a particular antimicrobial agent can no longer be used in the treatment of an infection.

3 *Metabolic mutants.* These mutants have lost the ability to use a particular carbon source and are usually affected either

in transport or metabolism. Other mutants which might be considered to fall into this category are mutants which have lost the capacity to make specific cell components, e.g. the capsular polysaccharide referred to earlier, and certain mutants which exhibit altered colonial shape, arising from mutations affecting the synthesis of the cell walls.

4 *Regulatory mutants*. The mutation in these organisms affects either the regulatory region of the promoter of the gene or the activity of a regulatory protein. The result is that the normal expression pattern of the operon or regulon is perturbed, e.g. a mutant which produces a defective *araC* gene product will be unable to grow on arabinose since the AraC protein is essential for the expression of the *ara* regulon. Such mutants were essential to the unravelling of the regulatory networks of bacteria.

These are but a very few of the mutations commonly encountered in nature and in the laboratory. In addition to classifying mutants in terms of their effect on the genotype of the cell, there are also different types of mutations. These are listed in Table 7.1 with a brief indication of their effect on the gene in which they occur. One type of mutant which deserves special mention is the *conditional* mutant. An example of this is a temperature-sensitive mutant (often shortened to *ts*) which is a special missense mutation which causes the protein product to be non-functional at high temperature. At low temperature the gene product is almost normal but at higher temperature is non-functional, allowing the effects of the loss of that gene product to be determined. Such mutations are isolated when the product of a gene is essential to the viability of the organism or virus and these too have been essential in understanding complex processes such as the assembly of viruses (see page 98).

Although mutation is a naturally occurring, spontaneous and undirected process, mutagens such as ultraviolet light, ionizing radiations and chemical substances (e.g. nitrous acid and nucleotide analogues) can be used to produce a

Table 7.1. Types of mutation

Type of mutation	Effect on gene product
Nonsense	Prevents completion of translation of gene product—truncated protein produced
Missense	Changes a single amino acid in the product—protein with reduced or altered activity produced
Insertion/deletion	*(a) Small:* causes complete mis-translation of the mRNA occurring after the mutation—produces protein with reduced activity *(b) Large:* removes part or all of gene—results in no gene product

generalized increase in the mutation rate. Such an increase is clearly valuable, but when the aim is to isolate a particular type of mutant, the difficulty is usually to find a suitable method to enrich specifically for the desired mutant and to recognize and isolate it when it may still be only a small proportion of the irradiated population. Direct selection can only be achieved when the mutation gives rise to resistance to a chemical or virus.

When more subtle mutations are required an enrichment step is usually used to increase the percentage of cells carrying the mutated gene. One such method is penicillin (or more frequently ampicillin) enrichment. Here the cells are initially grown in conditions which allow their growth at rates comparable to that of the parent cells. The mutagenized cells are then transferred to an environment in which the desired mutant cannot grow and after a short time (to allow growth of the mutant to cease) ampicillin is added to kill any cells which are still able to grow. The survivors of such a treatment are enriched for the desired mutant. A special method of transferring cells from one agar plate to another, *replica plating*, has been developed which makes the screening of populations of cells for the presence of mutants much easier (see Fig. 7.7 for an explanation of this technique).

The technique of replica plating was specifically developed to allow an important and controversial question

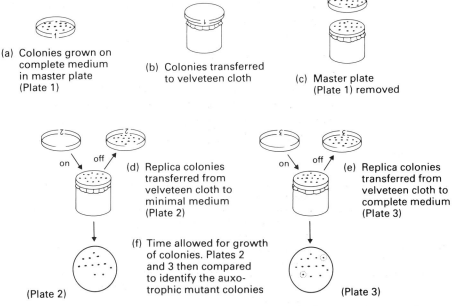

(a) Colonies grown on complete medium in master plate (Plate 1)

(b) Colonies transferred to velveteen cloth

(c) Master plate (Plate 1) removed

(d) Replica colonies transferred from velveteen cloth to minimal medium (Plate 2)

(e) Replica colonies transferred from velveteen cloth to complete medium (Plate 3)

(f) Time allowed for growth of colonies. Plates 2 and 3 then compared to identify the auxotrophic mutant colonies

(Plate 2)

(Plate 3)

Fig. 7.7. The technique of replica plating for transfering colonies from one plate to another. (Reproduced with permission of the American Society for Microbiology, Washington, USA).

to be answered. The question was whether mutations occurred spontaneously, or were directed by the selective agent. For example, consider mutation to penicillin resistance. Does penicillin direct this mutation or does it simply select out a naturally-occurring spontaneous mutant? Mathematical analysis of the occurrence of mutants showed that the process is indeed a spontaneous one. Further, replica plating allowed the isolation of penicillin resistant mutants without ever exposing the cells directly to penicillin.

Plasmids

One of the major sources of variation in the characteristics of cells within populations is due to the acquisition or loss of plasmids. Plasmids are small circular DNA molecules which replicate in cells independently of the chromosome. One of their most important characteristics is that the genes that plasmids carry are not essential to the cell but rather confer a novel property on that cell. They have been observed in most bacterial species and also in yeast. In addition to the genes for their own replication, many plasmids carry genes for other biochemical activities, e.g. toxin production, heavy metal resistance, metabolism of recalcitrant molecules, and resistance to antibiotics. It is in this last area that plasmids most impinge on our daily lives. Certain plasmids, known as R plasmids, carry the genes for resistance to many different antibiotics but in addition have genes which specify the proteins needed to transfer the plasmid to a new bacterium. It has frequently been observed that patients treated for long periods with antibiotics develop a population of resistant organisms, including species which are not normally resistant to antibiotics. This occurs by transfer of a resistance plasmid from one organism to another via conjugation (see below). One of the dangers of the existence and spread of plasmids carrying resistance to many different antibiotics is that sensitive organisms which are usually easy to treat with antibiotics can rapidly acquire resistance to a range of these drugs. Consequently, there have been attempts in recent years to control the use of antibiotics, particularly in agriculture were they have been used to promote better growth of pigs and chickens, for example.

On the plus side, plasmids lie at the core of genetic engineering which offers the potential to vastly extend the range of antibiotics available and to increase the availability of any compounds previously in short supply, e.g. insulin (see below).

114

Mutations and microbial adaptability

The selection of spontaneous mutants by the environment and the exchange of plasmids between organisms can occur very quickly in micro-organisms. This can lead to rapid changes in the genotype of the dominant organism in a population of cells. Although mutation rates are similar to those in higher organisms, various factors will lead to a faster rate of spread of a newly acquired characteristic, e.g.:

1   All prokaryotes and many eukaryotic micro-organisms are normally haploid so a mutation cannot be masked by the allelic gene.

2   They are unicellular, so any mutated cell can give rise to a new evolutionary line.

3   Micro-organisms can occur at high population densities in restricted environments giving a good chance of a mutant being present in a particular ecological niche where it may have a selective advantage.

4   Micro-organisms grow very rapidly.

5   Plasmid transfer between cells mean that new traits do not have to be acquired by all organisms by a long process of mutation and evolution.

As a result of these properties, micro-organisms, and, in particular, prokaryotes, have been able to adapt themselves to a wide variety of extreme environments (see Chapter 9). Another consequence, which can cause considerable embarrassment, is that they can also change rapidly during laboratory culture. The environment of a normal growth medium is often wildly different from that normally encountered in nature and a few sub-cultures may be sufficient to select for a series of mutants adapted to growth in this new environment. We have already seen an example in the *Pneumococcus* which loses its ability to produce a capsule and thus becomes avirulent (p. 106). Presumably the capsule protects the virulent organisms against phagocytosis in the animal host (p. 133).

It must be realized that a single bacterial colony may contain as many as 10^8 cells. A bacterium will have at least 10^3 genes and each will have a mutation frequency of about 1×10^{-8}. A single colony may then contain about a thousand mutants. In other words, a pure culture is a figment of the imagination, a very worrying thought to the microbiologist. Why do organisms not continually diverge? One major reason for this is that as the environment changes a mutation which previously had been either advantageous or neutral becomes deleterious. Consequently, cells carrying that mutation will be selected against. A further factor is that all mutated genes are subject to back mutation resulting in restoration of the original genotype. Thus, in a population an equilibrium exists for any given gene between forward and back mutation, and the character of the culture should remain relatively stable.

Although many of the changes that occur in the genotype may be minor and do not result in any obvious changes in the properties of the micro-organisms under study, they may be important if we wish to apply laboratory experiment results to events occurring in a natural environment. What can be done to prevent such selection occurring in the laboratory? All we can do is to choose environmental conditions to mimic as far as possible those occurring in nature and hope thereby to minimize change. An alternative is to avoid sub-culture by preserving the cells in a state of suspended animation by freeze-drying or, better still, by keeping them at liquid nitrogen temperatures. However, not all micro-organisms will survive such treatment.

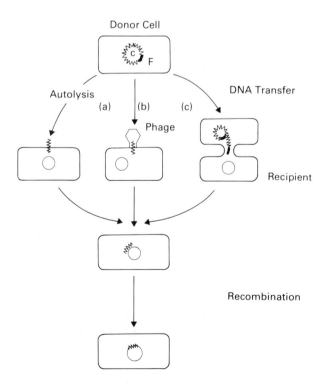

Fig. 7.8. Methods of transfer of DNA from one bacterium to another (a) transformation; (b) transduction and (c) conjugation.

Genetic recombination in micro-organisms

Recombination in prokaryotes

We have seen that genetic changes occurring as a result of mutation can result in the acquisition of new biological characteristics and thereby allow evolutionary change. However, evolution of the fittest organism in a particular environment will be aided if transfer of genes between one organism and another is made possible by genetic recombination. Compared with the ordered nature of sexual recombination in eukaryotes, the process in prokaryotes is less well developed, occurs rather spasmodically and does not involve a true fusion of male and female gametes to produce a diploid zygote. Instead there is transference of only some genes from the donor cell to produce a partial diploid. This event is followed by recombination to restore the haploid state (Fig. 7.8). There are three mechanisms by which these DNA fragments can pass from a donor to a recipient cell—transformation, transduction and conjugation.

1 *Transformation.* This process was referred to earlier (p. 106). Its discovery was one of the key experiments of molecular biology and as a technique it continues to occupy a central place in molecular biology. We have seen that virulence in pneumococci is determined by the ability to produce an extracellular capsular polysaccharide (p. 106). Indeed, pneumococci can be divided into about a hundred types which differ in the chemical nature of the capsular polysaccharides and hence in the specificity of the enzymes involved in their biosynthesis. It was found that it was possible to transform one type of penumococcus to another by adding an extract of the second type.

The substance in the extract responsible for transformation was purified and shown to be DNA. In the example given, the DNA of Type 2 cells had entered the Type 1 cells and the genes concerned in capsular polysaccharide biosynthesis had undergone recombination resulting in the insertion of Type 2 genes in place of the corresponding Type 1 genes (Fig. 7.9). The historical importance of the finding would be difficult to exaggerate since it showed clearly that DNA was the unit of heredity—in other words that genes are made of DNA. This was confirmed by the finding that other pneumococcal genes could be transferred in a similar way.

In transformable cells an important property appears to be a binding of the DNA to the cell surface prior to uptake. However, this property seems to be restricted to a few bacterial genera. This problem has been overcome to a

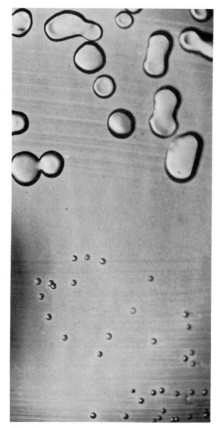

Fig. 7.9. Transformation of pneumococcal types.

significant extent in both prokaryotes and eukaryotes by the removal of the cell wall to form protoplasts. Addition of a compound (a fusinogen) which allows the protoplasts to fuse leads to the DNA of two or more cells becoming mixed and recombining to give new combinations of characteristics. This technique is known as *protoplast fusion*. When DNA is included in the incubation mixture transformation occurs simultaneously with fusion and thus small fragments of DNA, e.g. plasmids, can be introduced into cells. Protoplast fusion is not restricted to cells of the same species. Fusion between members of different species can be arranged by mixing protoplasts of the two species in the presence of a fusinogen. In this way genetic exchange can be achieved between organisms for which no other gene transfer system exists.

2   *Transduction.* This process is similar to transformation except that the DNA fragment passing from the donor to the recipient is incorporated into a phage. The donor cell is attacked by a phage, and in the process of virus maturation,

118

some bacterial genes are incorporated into the phage head either as part of an incomplete viral genome (*specialized transduction*, e.g. lambda phage) or the bacterial DNA replaces that of the bacteriophage in the phage head (*generalized transduction*, e.g. P1 phage). In specialized transduction the incomplete viral DNA, on entering the recipient cell, is normally unable to initiate the process of multiplication but instead either the phage enters the lysogenic state or the incorporated bacterial genes undergo recombination with the host DNA. If the transducing phage undergoes lysogeny then the recipient strain can become diploid for the genes carried by the phage. This property is very useful in genetic analysis. In generalized transduction there are no phage genes and so the only fate for the DNA is homologous recombination with chromosome of the recipient cell.

3 *Conjugation.* In this process cell contact is necessary. Transfer of DNA from the donor or 'male' cell to the recipient or 'female' cell involves special pili called 'sex pili' or F-pili (p. 27), which form a bridge between the cells. The DNA passes into the recipient F$^-$ cell in a unidirectional process such that an orderly entry of genes occurs according to their arrangement on the chromosome. Although in theory the whole chromosome can be transferred, the process is rarely complete in practice. An additional complication occurs in conjugation. Whilst in both transformation and transduction any cell can act as a donor or a recipient, in conjugation some cells are male and some are female in the sense that some are donors of DNA while others are recipients. What determines this primitive form of sex? Two experiments throw light on the problem.

(a) A female can be changed to a male by contact with a male, a process involving a passage of a plasmid present in the male called a 'sex factor' or 'F-plasmid'.

Fig. 7.10. Diagrammatic representation of the integration of the F plasmid to yield an Hfr strain and its excision to yield and F plasmid.

(b) A male (F+) can mutate to a 'supermale' which shows a much higher frequency of recombination with a female (hence the mutant is called Hfr). In such Hfr strains, the F-plasmid is integrated into the main chromosome where it duplicates in phase with nuclear division. During mating the

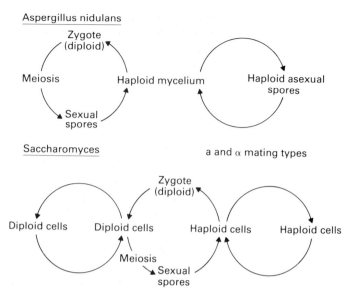

Fig. 7.11. Comparison of *Aspergillus nidulans* and *Saccharomyces* life cycles.

chromosome breaks at a point in the middle of the integrated F-plasmid and the entry of this piece of DNA into the female cell is followed by the chromosomal DNA. It is very rare for conjugation between an Hfr male and a female to lead to the female cells becoming male as the entry of the second half of the F-plasmid rarely occurs.

Occasionally, the integrated F-plasmid breaks free of the chromosome and in doing so takes with it a piece of chromosomal DNA. The resultant plasmid is called an F′ and is a useful way of transferring genes from one bacterium to another. These states are summarized in Fig. 7.10. Plasmids and bacteriophages which have the property of reversible integration into the host chromosome are called *episomes*.

The importance of recombination in prokaryotes

Are these recombination methods important in nature? Yes, indeed! Conjugation was discussed above as the principal means by which drug resistance spreads through a population of cells in patients undergoing antibiotic therapy. Transduction may also be common; most bacteria can be attacked by phages, many of which are capable of transduction. Furthermore, it must be realized that relatively rare processes may be of considerable importance in the evolution of new strains. What, then, has been the value of these methods in the laboratory? Here there is no doubt that they have proved invaluable in the development of molecular genetics; they have allowed genetic mapping not only of genes within a chromosome but also an understanding of the fine structure of genes and the relationship of this structure to the DNA molecule itself.

Recombination in eukaryotes

One of the major differences between prokaryotes and eukaryotes is that in the latter the genes are organized onto several chromosomes which are to be found in the nucleus. The number of chromosomes is species-specific and is a stable character of that species. Eukaryotic cells may carry two complete sets of the chromosomes and they are thus said to be *diploid*. Genetic recombination in eukaryotes is essentially a sexual process involving recombination during the process of meiosis leading to the eventual production of male and female haploid gametes. Fusion of such gametes results in the formation of new types of diploids. However, in the fungi an additional complexity is that, unlike most eukaryotes, some only exist in the haploid state. Consider the life cycles of two fungi, *Aspergillus nidulans* and *Saccharomyces cerevisiae* (Fig. 7.11).

In *A. nidulans* the growth form is a haploid mycelium (see p. 50). Mycelia from different strains can fuse and produce a fruiting body. In the fruiting body the nuclei of the two strains fuse to form a diploid nucleus which undergoes meiosis to reduce the number of chromosomes to the haploid state. During meiosis the chromosomes from the two nuclei pair up and recombine with each other to give hybrid chromosomes which carry genes from both parent chromosomes. Meiosis is followed by the formation of spores which subsequently germinate to produce new haploid hyphae (Fig. 7.11). The essential differences between this system and that in the yeast, *S. cerevisiae*, are that the organism does not normally form a mycelium, and, perhaps more importantly, the diploid formed from the fusion of the two cells can undergo normal growth and division. Thus, no fruiting body is formed and each of the diploid cells can undergo meiosis to produce haploid cells.

The alternation of a haploid and a diploid generation is characteristic of most eukaryotic micro-organisms, although there is a wide variation both in the relative importance of the haploid and diploid phases for vegetative reproduction and in the degree of sexual differentiation they exhibit. Thus, in fungi the predominant phase may be haploid or diploid or there may be a roughly equal role of both: morphologically

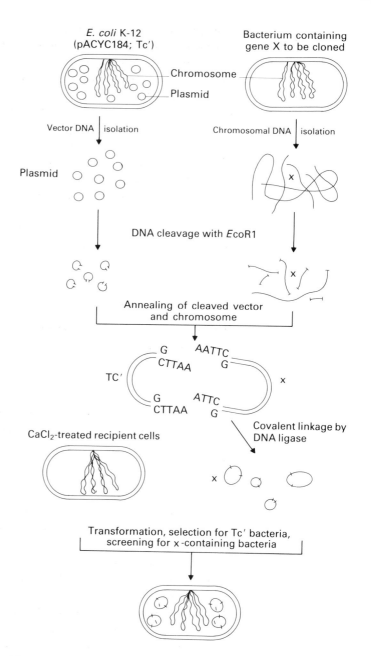

Fig. 7.12. The basic gene cloning procedure. (Reproduced with permission from Professor John E. Smith, *Aspects in Microbiology*, Vol. 11. Van Nostrand Reinhold (UK) Co. Ltd.)

distinct male and female forms can exist or there may be sexual conjugation between two similar cells derived from the same clone.

One important characteristic of eukaryotic genetics is the presence of extra-nuclear genetic DNA in organelles such as mitochondria and chloroplasts (see p. 31). These extra genomes encode essential components of the organelle in which they are found and thus interact with the nuclear chromosome to determine the characteristics of the organelle. These organelles also contain a separate transcription and translation system which is different from that in the cyto-plasm and which bears a greater resemblance to that of bacteria than that of eukaryotes. This characteristic is one of the major reasons for considering that organelles may have evolved from bacteria captured by the eukaryotic cell.

Eukaryotic cells may also possess plasmids. Thus, *S. cerevisiae* possesses a 2μ plasmid which has formed the basis of gene cloning in yeast both in its own right and as a hybrid molecule with bacterial plasmids allowing the analysis of yeast genes both in bacteria and in the yeast itself.

Genetic engineering

The importance of genetic exchange between organisms was described above. The methods of achieving gene transfer which have been described so far have largely been naturally occurring processes adapted to suit the needs of the experi-mentalist. In the last ten years there have been major new developments which allow the transfer of genetic information between species, and indeed kingdoms. This technique is called *genetic engineering* and rests on two of the major discoveries of the last twenty years, namely, *plasmids* and *restriction enzymes*. Plasmids were described above and their general properties will not be discussed again.

Restriction enzymes are one of the bacterial defence systems against entry of foreign DNA. The restriction enzyme recognizes a specific sequence of bases in the DNA and cuts the DNA at or close to that site. This specific sequence of bases occurs within the DNA of the bacterium but is pro-tected from the action of the restriction enzyme by modification of one of the bases within the sequence. Foreign DNA is not similarly protected so that when it enters the cell it is rapidly degraded by the restriction enzyme. Thus, DNA which can survive entry into the cell is limited to that from organisms with the same restriction/modification system.

The plasmids used in genetic engineering are relatives of the naturally occurring plasmids found in micro-organisms. Generally, they have been mutated and tailored with enzymes such that their properties reflect those required for genetic engineering. For example, drug resistance genes are intro-duced to allow easy identification of cells carrying the plasmids, the number of sites at which a restriction enzyme cuts is reduced to one to allow insertion of 'foreign DNA', and control systems to allow the controlled expression of the gene of interest have been introduced. The advantage of genetic engineering over conventional genetic techniques is that one can transfer a single specific gene between organisms whereas conventional techniques transfer hundreds of genes simultaneously. The steps to achieving this are as follows:

1   The donor DNA carrying the gene of interest is cut with a restriction enzyme to yield fragments of various sizes, one of which bears the gene of interest. A suitable plasmid cut with the same enzyme is mixed with the donor DNA and the two are joined by an enzyme, DNA ligase, to give a series of hybrid plasmids.

2   The hybrid plasmid DNA is used to transform the host cell and the cells are plated out onto agar. The ratio of the amount of DNA to cells is adjusted such that each cell only takes up one DNA molecule, thus each colony which grows represents a different piece of the donor DNA carried on a plasmid. One of these carries the gene of interest and can be recognized by the characteristic it confers on the cell. This gene is then said to have been *cloned* (Fig. 7.12).

This then is the basis of genetic engineering. There are a great many subtleties which cannot be described here. However, the method can be applied to DNA from any organisms and the host can be a bacterium, yeast, a fungus, or animal or plant cell. Some of the successes of genetic engineering in the industrial sector will be described in Chapter 10, but these techniques have also been responsible for illuminating our understanding of the molecular mechanisms of adaptability in micro-organisms.

Apart from the obvious advantages of genetic engineering to the microbiologist it has been recognized that there are dangers associated with this technology too. Thus, the possibility of creating new and uncontrollable pathogens from tame micro-organisms such as *Escherichia coli* and yeast has caused genuine concern in the world at large. To counter these possibilities, certain classes of genetic engineering can only be carried out in special laboratories where the new organisms can be 'contained'. In this way our understanding of the molecular basis of pathogenesis and cancer-causing viruses can be safely advanced.

8 : Associations between Micro-organisms and Higher Organisms

Micro-organisms have many effects on man, some direct, some indirect, some beneficial and some harmful. As a result of Koch's demonstration in the last century of the ability of specific microbes to cause specific diseases to man, microbiology has been dominated by focusing on their harmful effects; it is only recently that more attention has been paid to their positive value. This chapter and the next two will give some illustration of the many ways in which microbes can affect us.

It is common in nature for micro-organisms to be associated in some way with a host. Micro-organisms can form such relationships with other micro-organisms, with animals or with plants and as a result may gain protection, a useful source of nutrients or other benefits. The host may be unaffected, in which case the relationship is called *neutralism*, it may gain (i.e. *mutualism*) or it may lose (i.e. *parasitism*). However, the following points should be noted concerning such relationships:

1   It can be argued that in a close relationship of the type implied, the host cannot be unaffected and hence the term neutralism has little meaning other than in a relative sense.

2   The borderline between mutualism and parasitism may not always be clear. The term *'symbiosis'* is best used in its original meaning of 'living together' and is useful to describe relationships without making judgements on gains or losses between the partners. Symbiosis can also be divided into endosymbiosis in which one organism lives within the cells or tissues of the host and ectosymbiosis in which the symbionts live separately.

3   It must be emphasized that the advantage gained by the micro-organism from an association with a host must be

considered in terms of the population rather than the
individual.

Neutralism, mutualism and parasitism

In order to illustrate the gradation from an apparently neutral
relationship to a mutualistic or a parasitic one, let us consider
three systems—the mammalian gut, the human mouth, and
the plant root.

Micro-organisms and the mammalian gut

The majority of micro-organisms in the gut occur in the small
and large intestines, the acidity of the stomach being sufficient
to prevent the growth and to cause the death of micro-
organisms. A few acid-tolerant bacteria can colonize the
lining of the stomach but their effect on the host is unknown.
As we pass from the small to the large intestine and to the
bowels, the concentration of micro-organisms increases so
that they make up about a third of the weight of faeces. In
some ways, therefore, the gut resembles a continuous culture
system with an entry port in the mouth and an exit port at the
anus.

What are these micro-organisms and what are their
effects? Surprisingly enough the majority of these have only
been grown in the laboratory in recent years and it used to be
thought that most were facultative anaerobes like *Escherichia
coli*. Indeed, before its contemporary claim to fame, the main
importance of *E. coli* to the microbiologist was as a measure
of the faecal contamination of drinking water, the number of
organisms in a unit volume giving a reasonable indication of
the contamination. However, it seems that most gut
organisms are very strict anaerobes which are killed by even
the small amounts of oxygen normally used in laboratory
sub-culture and special methods are necessary for their
isolation. In the intestines these strict anaerobes are reliant on
the facultative anaerobes to remove oxygen from the
environment.

Although we now know the nature of the normal gut
flora, their effect on the animal host is unclear. It used to be
thought that the micro-organisms were essentially neutral but
more recent evidence renders this unlikely, and suggests that
they are mutualistic symbionts. It is possible to maintain
animals which are completely free of micro-organisms, but
such 'germ-free' creatures have poorly developed defence
mechanisms, have an abnormal gut and suffer increased
nutritional needs compared with normal animals. Some of

126

these effects can also be produced in humans who are under-going medical treatment using wide-spectrum antibiotics; the majority of the gut flora are killed and as a result any normally harmless micro-organisms remaining may become dangerous. Evidently our gut bacteria perform a useful but not irreplaceable function and we should avoid their removal.

A more obvious example of mutualism is in the gut of ruminants, those herbivorous animals such as cattle, sheep and goats which use plant cellulose as the major carbohydrate source of their diet. Cellulose cannot be digested in a normal gut, but the ruminants have developed a special region for the purpose—the rumen. This resembles a large fermentation vat (about 100 litres in the cow) into which masticated plant materials enter to be digested by large numbers of anaerobic bacteria and protozoa. These mutualistic micro-organisms hydrolyse cellulose and other plant polysaccharides to their component monosaccharides which are then fermented to simple fatty acids and to methane and carbon dioxide. The fatty acids are absorbed from the rumen and are used by the animals as a carbon and energy source; the microbial cells produced at this time pass into the stomach where they are digested in the normal way and so provide the other nutrients (e.g. vitamins and much of the protein) required for growth. The process is summarized in Fig. 8.1. The similarity of the rumen to a continuous culture fermentation is striking and chemostats have recently been built in the laboratory to simulate its action.

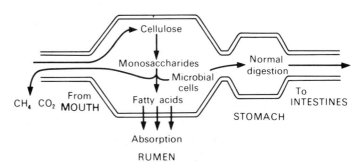

Fig. 8.1. A diagrammatic representation of the metabolism occurring in the rumen.

Micro-organisms in the mouth

Just as we have seen for the gut, the oral cavity has its own complex and characteristic microflora. This has a special place in microbiology, for it was from scrapings from his teeth that Leeuwenhoek so dramatically gave us the first descriptions of bacteria (see Fig. 1.2). He estimated that he had more bacteria living on his teeth than there were people in Holland. He clearly recognized five types of bacteria, and we know today the normal oral flora has a notably high species diversity. It is composed chiefly of Gram-positive cocci, rods and filaments, but aerobic and anaerobic Gram-negative cocci and rods can always be found as well (see p. 13).

These communities of micro-organisms can live benignly in the mouth in neutralism, but given the right conditions they can develop to cause dental disease, for example caries, the localized dissolution of tooth enamel by acid. The major organism implicated in caries is *Streptococcus mutans*. This bacterium utilizes dietary sucrose in two ways: to make extracellular polysaccharides which aid in its adherence to the tooth enamel; and in fermentation, producing lactic acid. Thus, it can set up localized sites of attack on the enamel. It is clear, however, that caries, unlike other bacterial diseases, is the result of complex interactions between many different organisms living with and interacting with the *Streptococcus*. It can be prevented by good oral hygiene and a lowered sugar intake.

Micro-organisms and the plant root

In general, healthy plant leaves and stems do not provide a suitable environment for a large amount of microbial growth, but they do have their own characteristic microfloras. The roots, however, are surrounded by a region called the rhizosphere where microbial nutrients are provided by excretion from the plant and where the environmental conditions do not fluctuate so much as above ground. Consequently, the number of micro-organisms in this region may be many times greater than in the soil away from the root influence. The action of such rhizosphere organisms is not entirely clear but normally they appear to be neutral. They may play an important role in antagonizing the growth of potentially pathogenic organisms. However, a more organized relationship can exist with fungi in which the hyphae form an external sheath around the root and may even penetrate it. These mycorrhizas, as they are called, confer an advantage to plants growing in poor soils where they aid in the absorption of essential nutrients, especially phosphate, by increasing the area for absorption. The growth of forest trees is particularly improved by such a relationship.

Two contrasting relationships between bacteria and plant roots exemplify the spectrum of interactions that we find in nature. *Rhizobium* and *Agrobacterium* are closely related Gram-negative bacteria (they show very close DNA homology). Both infect plant roots to cause abnormal localized growths, but as R.A. Lewin commented, 'lodgers who pay rent are symbionts, whereas free-loaders who do not are parasites'.

The relationship between *Rhizobium* and leguminous plants (beans, clover, peas etc.) gives us a striking example of a well-developed mutualistic symbiosis. The rhizobia infect the plant via the root hairs and invade the tissues. Some of the plant cells are infected intracellularly and as a result enlarge, divide and eventually produce characteristic nodules, each of which may contain as many as 10^9 bacteria (Fig. 8.2). The importance of the relationship lies in the ability of *Rhizobium* cells in the nodule to fix atmospheric nitrogen. Such symbiotic nitrogen fixation is an exceedingly important process in agriculture since the limiting factor for plant growth is often the level of a suitable combined nitrogen source. A crop of leguminous plants such as clover may fix as much as 350 kg of nitrogen per hectare in a season compared with only about 5 kg per hectare for free-living nitrogen-fixing bacteria in temperate regions of the world or up to 60 kg per hectare by blue-green bacteria in the tropics. The importance of nitrogen fixation to the nitrogen cycle as a whole is discussed in the next chapter.

In contrast, *Agrobacterium* gives us an example of para-sitism, by forming crown gall tumours. The bacteria enter the plant through a wound and move between its cells, which are stimulated to rapid division. There is much current interest in this process as the bacteria have a large plasmid, the Ti (tumour inducing) plasmid, part of which is transferred to the plant cell, and incorporated and expressed in the plant nuclear DNA. Thus, it is possible that this plasmid could be

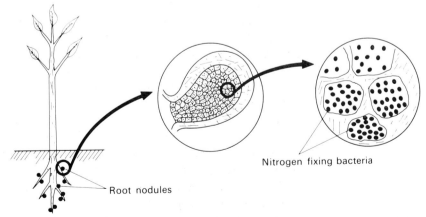

Nitrogen fixing bacteria

Root nodules

Fig. 8.2. Symbiotic nitrogen fixation showing a plant with root nodules and two magnifications of cross section through a nodule.

129

used to transfer genes to plant cells to produce genetically engineered plants.

Mutualistic associations between micro-organisms

The best known examples of mutualistic relationships between micro-organisms are the lichens in which fungi live in association with algae. Lichens are common in situations in which other forms of life are unable to survive, for example exposed rock surfaces and tree trunks.

The association is one in which both partners become modified to give a complex structure in which the bulk is made up from fungal hyphae that surround and penetrate a layer of algae (Fig. 8.3). Both symbionts can be grown separately in the laboratory, but it is very difficult to re-synthesize the lichen under these conditions. The symbiotic relationship in lichens is one in which the alga supplies the fungus with organic nutrients produced by photosynthesis while the fungus supplies the alga with a protective environment. The lichen itself has a marked ability to scavenge and to concentrate the limited nutrients available, an ability that makes it a sensitive indicator of air pollution, such as that involving sulphur dioxide. Most species are rapidly killed as the concentration of toxic pollutant rises and therefore the lichens cannot grow in industrial areas.

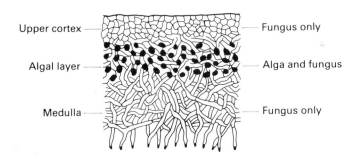

Fig. 8.3. The structure of a lichen seen in section.

Parasitism

In any parasitic relationship between a micro-organism and a higher organism, a cycle occurs. The parasite enters the host usually at some specific entry point; this entry is followed by consolidation and multiplication within the host and finally by dissemination, usually again at some specific exit point to allow infection of other hosts. In considering the factors in this relationship, we will be taking examples mainly from

130

animal infections since more is known about them and the means of defence against them. This selection must in no sense minimize the importance of the many other parasitic relationships that can occur and which have great economic and ecological impact.

1   *Entry into the host.* Animals and plants have adapted outer surfaces to provide a barrier to invasion which is normally broken only by damage or by means of a special entry mechanism such as by an insect bite. However, there are often particular areas on the host that are less protected and can be more readily invaded. Micro-organisms parasitic on animals are usually specialized to one of these areas such as the gut, the respiratory tract or the urino–genital system and those on plants to stomata or lenticels.

2   *Multiplication.* Once the micro-organism reaches a suitable region of the host and provided it has not been killed by the host defence mechanisms, multiplication occurs. The *pathogenicity* of an organism represents the ability of a parasite to gain entry and multiply, eventually causing the physiological and anatomical changes characteristic of a disease. This pathogenicity is usually very specific in terms of the host attacked. Thus, most diseases affecting domestic animals are fortunately not transmitted to man and vice versa. The few that are transmitted may cause a much milder form of the disease, a fact made use of in vaccination against smallpox by using a related virus attacking cows (cowpox). The degree of pathogenicity is termed the *virulence* of a micro-organism. A highly virulent strain may require only a single cell to cause an infection whereas others may need the entry of a large number of cells. Whilst pathogenicity is often used to define a species, variations in virulence commonly occur within the strains of a species. It is also possible to obtain mutants varying in their virulence (see Chapter 6) and such mutants are useful in defining the physical basis of virulence by comparison with the properties of the wild type. They are also valuable since avirulent mutants can induce a resistance to the disease without causing any dangerous symptoms and thus they can be used as live vaccines.

The multiplication of a pathogen is usually confined to certain specific areas of the host. Thus, the organism causing the disease of brucellosis in cattle grows preferentially in the foetus and the placenta of a pregnant cow, usually resulting in abortion. The reason for this localization within the host is that the growth of these bacteria is markedly stimulated by the presence of erythritol, a sugar absent from the tissues of the cow except for those specifically associated with the foetus. However, in most cases we do not know the reason for localization and further research in this area may well provide clues to aid the combat of an infection.

The symptoms of a disease are not usually caused by the simple mass of cells produced, although very large numbers may sometimes occur (as in the lungs of a patient infected by pneumonia). More commonly, it is the *products* of the pathogens—the *toxins*—that cause the symptoms. These substances are usually produced extracellularly, although in Gram-negative bacteria, the cell wall lipopolysaccharides are often toxic. The toxins have a range of actions. Some are enzymes; for example they may lyse with red blood cells (the haemolysins) or bind specifically to essential cell components as instanced by the diphtheria toxin which inhibits the transfer of amino acids from RNA to the growing polypeptide chain in protein synthesis. Microbial toxins act in extremely small amounts but are amongst the most lethal compounds known. For example, 1 mg of botulinus toxin is sufficient to kill more than a million guinea pigs. It is evident that a massive infection is not necessarily required to cause death; unfortunately, it is also obvious that such toxins pose a potential threat if they are to be used as the agents of biological warfare.

The growth of the microbial pathogen may result in the death of the host, although from the point of view of an efficient parasite in a long-term ecological situation, it is obviously advantageous to set up an equilibrium by which relatively little damage is done to the host which continues to live and release parasites to colonize fresh hosts. For example, a human typhoid carrier may have no recognizable symptoms yet will go on excreting typhoid bacteria for a long period of time, thus posing considerable problems for the epidemiologist concerned in the eradication of the disease.

3 *Release.* Micro-organisms are released from the infected host. This commonly occurs in a particular region which is characteristic of the disease and may or may not be the same as the point of entry and multiplication. In animals, release can occur from the respiratory system as droplets (coughs and sneezes), the alimentary tract usually in the faeces, the urinogenital tract or the skin.

4 *Spread.* There are many ways in which a micro-organism may spread from one host to another. Direct contact may be involved as in venereal disease. Spread may be airborne carried in droplets, dust or spores as in most respiratory infections or in fungal diseases of plants. Many intestinal infections are water-borne or are spread through infected food. Alternatively, a carrier such as an insect may be concerned. The mode used will depend partly on the ability of the

micro-organism to survive outside the host. An example of a short survival time can be seen in the bacterium causing the venereal disease of syphilis, which requires direct contact for transmission. At the other extreme are sporing organisms which may have very long survival times. The island of Gruinard off the Scottish coast was experimentally infected with anthrax spores in 1943 and it is still too dangerous to land there without special precautions. The spread of infection may involve a further multiplication; for example, in food or in an insect vector. Indeed, in some organisms the normal means of multiplication is outside the host which then becomes infected almost by chance.

In order to eradicate a particular disease we can interfere with any of these four phases in the cycle. Unfortunately, it is outside the scope of this volume to deal with all these phases in detail and further discussions must be restricted to events occurring within the host.

Resistance to infection

The tissues and fluids of animals, and to a lesser extent of plants, provide an ideal nutrient environment for microbial growth, and such growth can be very rapid if allowed to proceed unhindered. How have we and other higher organisms managed to survive and evolve in the light of the potential strength of invasion exhibited by micro-organisms? The reason, of course, lies in our ability to combat infection. Although the means used are, strictly speaking, a characteristic of the host rather than of the micro-organism, their study has usually been coupled with that of microbiology, for historical reasons. Resistance to infection can take two forms; it can be of a non-specific type present in hosts not previously exposed to infection—a constitutive resistance, or it can be induced by the presence of a particular micro-organism.

1 Constitutive resistance

It has been said that the surface layers of a plant or an animal act as a primary barrier to microbial invasion. However, if entry is effected, then there is a variety of defence mechanisms open to the host. Non-specific antimicrobial agents may be produced; one of these is the enzyme lysozyme (see Chapter 2), present in natural secretions and extracts of animal organs, while plants often produce phenolic antimicrobial agents. However, the most important constitutive defence mechanism in animals is *phagocytosis*. A whole series of different types of phagocytes present a defensive system, starting with the macrophages at the primary focus of infection, continuing with the polymorphonuclear leucocytes (which are attracted to the regional lymph nodes by the inflammatory response to the white blood cells), and ending with the reticuloendothelial system. Phagocytosis is normally followed by the lysis of the engulfed cells. However, a few microbes are able to avoid this fate; some are resistant to phagocytosis (e.g. the capsulate pneumococcus discussed in Chapter 7); some are engulfed but are unaffected by the lytic enzymes of the phagocyte and may, as in the case of the tubercle bacillus, actually grow intracellularly; others produce substances called leucocidins which kill the phagocytes. A quite different defence mechanism occurs against virus attack; infected cells produce a protein called *interferon* which prevents further viral attack in a non-specific manner.

It is evident that there are various levels of defence and that micro-organisms differ in their ability to overcome them—a measure of their pathogenicity and virulence.

2 Inducible resistance-immunology

In higher animals there is a secondary defence against infection should a particular organism succeed in establishing itself. The micro-organism induces the formation of specific proteins called *antibodies* which combine with them and render them more susceptible to lysis or phagocytosis. This production of specific antibodies is part of a general phenomenon by which animals react to the presence of foreign substances called *antigens* and it is not unlike the stimulation of induced enzymes by the inducer. The study of the immune response is called immunology and it is of considerable importance both to medicine and to biology in general. Let us consider the nature of antigens and antibodies and the consequences of the reaction between them.

Antigens. All sorts of substances are capable of inducing the synthesis of specific antibodies—in other words are capable of acting as antigens. In general, they have to be high molecular weight polymers such as proteins and polysaccharides. On the surface of such macromolecules there are particular regions called *antigenic-determinant groups* that react with antibodies; a protein molecule may have as many as a hundred of such groupings corresponding to particular arrangements of amino acids. On the other hand, low-molecular weight substances are not normally antigenic, although they may become determinant groups when combined into a large molecule. The most important characteristic of this antigenic capacity is in the specificity of the antibody produced. Protein molecules that differ only in one or two amino acids along the chain or chains can be distinguished. Indeed, the specificity is similar to that between an enzyme and a substrate and implies a non-covalent linkage between the antibody and the antigen which depends upon a close approximation of complementary surfaces akin to a lock and key.

Antibodies. Antibodies are proteins occurring in the globulin fraction of serum. Since virtually every protein and many polysaccharides injected into an animal can evoke the formation of a unique antibody and since many substances that have only been produced artificially in the laboratory can be made to act as antigenic-determinant groups, a very large number of different antibodies can potentially be produced by a single animal, a number that has been estimated as being at least 10^6. The antibody molecules are made up of two specific regions joined by a non-specific region, this latter area shared by a whole group of antibodies. The specific regions are the ones which actually combine with the antigen so that the antibody behaves as a divalent molecule, an important characteristic in the formation of precipitates or in agglutination. The specific regions are different in each different antibody, i.e. are variable, and are made of two polypeptide chains (Fig. 8.4).

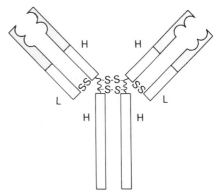

Fig. 8.4. Diagram of an antibody molecule. It is made of four polypeptides, two heavy and two light, held together by disulphide bridges. The shaded portions are the variable regions which bind the antigen between them.

One of the puzzling aspects of immunology has concerned the capacity of animals to produce such a wide range of different protein molecules, albeit with certain structural similarities. The answer involves some unusual genetic mechanisms in the B-lymphocytes that make the antibodies. Each individual B-cell makes only one specific antibody, as during its development its pool of genes for each of the two variable parts of the antibody molecule undergo site-specific recombination events to yield one of thousands of different possible polypeptides. These two can thus associate to yield one of millions of different antibody molecules. Even more variation is probably built into this system by allowing for somatic mutations to occur in these genes coding for the variable regions. If triggered by its particular antigen, the B-cell multiples rapidly and each cell enlarges to become a plasma cell actively secreting the specific antibody that has been triggered. This mechanism explains the fact that once we have been exposed to a particular infection, we gain a specific immunity since an excess of the specific antibody-forming cells will remain.

The antibody-antigen reaction

The basic reaction is a combination of specific regions of the antibody with the appropriate determinant groups on the antigen. Since the antibody is divalent and the antigen is normally multivalent, such a reaction leads to the formation of a lattice-like multimolecular complex (Fig. 8.5). If the antigen is soluble, precipitation will result.

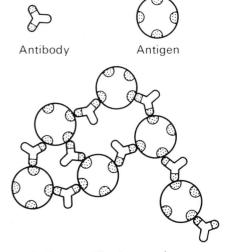

Antibody   Antigen

Antigen-antibody complex

Fig. 8.5. The formation of a precipitate between a divalent antibody and a quadrivalent antigen (the combining sites are represented by shaded areas). The lattice must be considered as growing indefinitely in three dimensions.

Let us consider what will happen when an animal is infected with a micro-organism to which it has not been previously exposed. The micro-organisms are phagocytosed and their components eventually find their way to the lymphoid tissues where antibodies are produced against them. However, we have seen that a micro-organism will contain about a thousand different protein molecules, as well as polysaccharides and other potential antigens. In theory, antibodies will be formed against all of them, although in practice surface antigens are the most important. The polymers of the cell wall, the capsule, pili and flagella, occur on the outer surface of the cell and are usually present in large amounts compared with other antigens. We can consider the cell as one vast macromolecular antigen with the determinant groups on these surface polymers. Reaction with the appropriate antibody leads to a cross-linked lattice resulting in the precipitation or rather the agglutination of the cells. In the body, such a combination with antibody leads to a much increased susceptibility to the host defence mechanisms and lysis and phagocytosis eventually result.

The most important use of immunology to the microbiologist lies in its provision of a means of identification by the use of an appropriate specific antibody. For example, we have seen that there are over a hundred types of pneumococci which vary in the chemical nature of the capsular polysaccharides and we can distinguish between them by testing the ability of specific antibody preparations to agglutinate our unknown pneumococcal suspensions. Such testing is of particular value to epidemiologists trying to trace the spread of an infection as it enables them to 'fingerprint' a specific type and study its distribution and movement.

Monoclonal antibodies. Antibodies are very useful in the laboratory, for example in the diagnosis of disease. From the above discussions it is clear that the blood of an animal will contain a myriad of different antibodies, from which it is very tedious to purify one particular antibody of choice. It is now possible to manufacture this particular antibody at will by using a specific cell line. To do this a mouse is immunized with, for example, a virus preparation, so that some of its B-cells will make the antibody. B-cells, however, cannot be grown in culture. Thus, a preparation of the B-cells is fused with 'immortal' cells derived from a lymphocyte tumour. The resulting *hybridomas* can then be grown indefinitely as *clones* from individual cells. The clones are tested for the desired antibody, and a positive one can then be propagated indefinitely to give its *monoclonal antibody.*

Antimicrobial agents and chemotherapy

As soon as it was realized that disease could be caused by micro-organisms, attempts were made to produce chemicals which killed them, or at least prevented their growth, and so the search for antimicrobial agents was initiated. At first, cheap substances which killed as many microbes as possible and could be used for chemical sterilization were sought. The most effective compounds were substances like chlorine, hypochlorites, phenol, salts of heavy metals and detergents. Those agents called *disinfectants* were used on inanimate objects or, in treatment of water, at low concentrations so that possible toxicity to higher organisms was not important. Indeed, most of the chemicals were general protoplasmic poisons acting, for example, by denaturing proteins. Soon people became concerned with the possibility of producing substances which could also be used in contact with higher organisms and, in particular, with man. *Antiseptics* were required for skin disinfection or in the treatment of small wounds, while *chemotherapeutic agents* were designed to act on already established infections through their absorption into the circulating fluids. For this purpose substances were required which had a specific inhibitory effect against micro-organisms—that is, had a *selective toxicity*. None of the substances commonly used as disinfectants had this property since unfortunately higher organisms are generally much more susceptible to chemicals than are micro-organisms and, even more so, bacteria and viruses. Although Lister had introduced the use of phenol in his 'aseptic' surgery and many of the results proved beneficial, the introduction of phenol into open wounds was almost certainly harmful since it inhibited the host defence mechanisms as a result of its selective toxicity towards animal cells compared with micro-organisms. A selective toxicity to micro-organisms was required but how could such substances be discovered?

In the early years of this century the search for such an agent was inspired by Ehrlich's concept of a 'magic bullet' having such selective killing power. This involved the testing of a wide variety of synthetic organic compounds, and resulted in the discovery of Salvarsan, an arsenical compound lethal for the spirochaetal organisms of syphilis. After this, no real progress was made until the discovery of sulphonamides, a group of chemicals based on the structure of sulph-anilamide. Although the sulphonamides were and still are used against certain bacterial infections of animals and men, their discovery initiated a more general search for agents with selective toxicity. The reason for this arose from the finding that their inhibitory effect against sensitive bacteria could be overcome by adding *p*-aminobenzoic acid. Closer study showed that the sulphonamides acted by competing with *p*-aminobenzoic acid for the active site of an enzyme which converts the latter to folic acid, an essential coenzyme for a variety of metabolic conversions.

This competitive inhibition was due to the similar structure of sulphanilamide and *p*-aminobenzoic acid, a general phen-

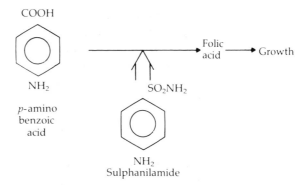

COOH

NH₂

p-amino
benzoic
acid

SO₂NH₂

NH₂
Sulphanilamide

Folic acid → Growth

omenon well recognized in biochemistry. This discovery led to great excitement since it opened up the possibility of a systematic production of chemotherapeutic agents by synthesizing analogues of microbial metabolites other than *p*-aminobenzoic acid.

Microbial metabolite ────→ Growth

Analogue of microbial metabolite

Since then, a very large number of analogues have been produced, many of which inhibit microbial growth. However, most of the compounds are toxic to higher organisms for the simple reason that most microbial metabolites are common to all the cellular forms of life because of the unity of biochemistry referred to previously. Substances and reactions must be found which are specific to micro-organisms and inhibitors produced against them. However, this approach has unfortunately still not led to the production of any really effective chemotherapeutic agents, but the reason may lie partly in lack of knowledge of the more sophisticated mechanisms involved in the growth and metabolism of micro-organisms.

Before long a new approach began to dominate chemotherapy, stemming from the discovery that micro-organisms can themselves produce antimicrobial agents called *antibiotics*. This phenomenon had been known since the early days of microbiology; in 1929 Fleming had described an antibacterial agent produced by the fungus *Penicillium notatum* which he called penicillin and which seemed to have many of the properties required for an ideal chemotherapeutic agent against Gram +ve bacteria (Fig. 8.6). However, little immediate progress was made in its purification. In those days, chemists knew little microbiology,

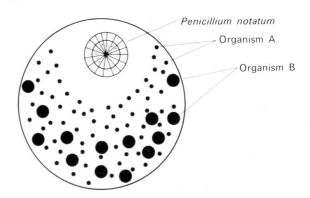

Fig. 8.6. The results of an experiment similar to that by which Fleming originally discovered penicillin. A plate is seeded over the whole surface with two organisms, one producing small colonies (A) and one producing large colonies (B). Penicillin diffuses from the colony of *Penicillin notatum* and inhibits organism B at lower concentration than organism A.

and microbiologists knew little chemistry; there was almost no understanding between the two—biochemistry was in its infancy. So it was not until the early 1940s that methods were finally perfected by Florey and Chain for the purification of penicillin and it became clear that the antibiotic was a far more effective chemotherapeutic agent than anything that had been previously discovered. Suddenly it was realized that antibiotic-producing microbes were not uncommon, particularly in soil where they are thought to prevent the growth of competitors. Microbiologists came back from their holidays weighed down with soil samples and with the gleam of Nobel Prizes in their eyes! A large range of antibiotics was discovered and chemotherapy was revolutionized to the extent that most infections caused by cellular micro-organisms could be treated.

What, then, is the basis of the selective toxicity of these antibiotics? This is not just an academic question since more and more micro-organisms are developing resistance to the substances in use today (see Chapter 7). A knowledge of the mode of action of successful antibiotics should provide clues to the design of new and possibly better chemotherapeutic agents. Such a knowledge is slowly becoming available—slowly because it requires a detailed understanding of microbial physiology. For example, in spite of a vast number of papers published on the mode of action of penicillin, it is only recently that the activity has been pinned down as an inhibition of the final stage of the biosynthesis of the peptido-glycan of the prokaryotic cell wall. At once, the selective toxicity of penicillin was explained since eukaryotes, like

140

ourselves, have no structures resembling the peptidog¹
The sites of action of some important antibiotics are ind
in Table 8.1.

Table 8.1. Sites of action of some antibiotics

Antibiotic	Site of action	Effect
Penicillin ⎫ Cephalosporin ⎭	Cross-linking of peptidoglycan	Inhibits bacterial cell wall synthesis
Streptomycin Chloramphenicol Erythromycin	30s ribosome subunit Translation of mRNA on 50s ribosome 50s ribosome	Inhibits bacterial protein synthesis
Rifampicin	RNA polymerase	Inhibits mRNA synthesis
Nalidixic acid	DNA gyrase	Inhibits DNA replication
Polymixin	Cell membrane	Destroys membrane functions

Although selective agents against cellular micro-organisms and, in particular, against prokaryotes, are now available, the same is only just becoming true for agents against viruses. Unfortunately, the multiplication of the virus is entirely dependent upon the host cell and any attempt to inhibit this multiplication almost inevitably results in the destruction of the host cell as well. Nevertheless, some agents are becoming available with antiviral specificity, especially synthetic drugs interfering with viral nucleic acid metabolism. An example is acylclovir, a directive of guanine which can be used to treat herpes infections.

There is also active development of chemotherapeutic agents against cancer, including a range of cytotoxic (cell-killing) antibiotics produced by bacteria. For both antiviral and anticancer therapies there is currently great hope for the successful utilization of industrially produced interferon (see p. 171).

Microbiological control

In general, parasitic micro-organisms do harm to man, either by causing disease directly or affecting animals and crops. However, they can also be useful, for example in the elimination or prevention of the spread of pests. Consider myxomatosis, a virus disease of rabbits. The disease occurred in the western hemisphere and rabbits were generally resistant to it in such areas. In 1953, the disease was introduced to Europe where it had previously been absent. There was a catastrophic epidemic amongst the sensitive rabbit population which was almost wiped out. Agricultural land became available which had previously been ravaged and although there has been a later tendency for resistant rabbits and less virulent viruses to develop, the potentiality of such microbiological control is obvious. In particular, there has been much interest in the eradication of insect pests by this method, an attractive idea since it is not associated with the pollution dangers that attend the use of chemical insecticides. Three types of insect pathogens are used: the bacterium *Bacillus thuringiensis* which produces a potent toxin, fungi which infect insects on contact, and viruses.

9 : The Ecology of Micro-organisms

Micro-organisms make up a significant percentage of the total biomass on the earth. Indeed, as a result of their high rate of growth, their metabolic activity, and their adaptability, they are of even more importance in the proper functioning of the ecosystem than their total mass would indicate. They are also ubiquitous. Any environment supporting higher organisms will contain micro-organisms while the converse is not true, and their absence from an environment indicates special and unusual conditions. This chapter will consider the reasons for this ubiquity, the properties of natural environments as habitats for micro-organisms and their role in maintaining environments on the earth suitable for life and living as we know it.

Effects of environmental factors on microbial growth

In Chapter 5 we discussed the growth of micro-organisms and saw that it could be very rapid. Maintenance of high rates of growth requires careful control of various features of the physical and chemical nature of the environment. The most basic of these is provision of nutrients from which the micro-organisms obtain their raw materials and energy. In addition, the physico-chemical conditions must be suitable and in natural environments determine, along with other factors, the rates of microbial growth and the nature and size of the indigenous population.

Temperature

One of the most important factors affecting the rate of microbial growth is the environmental temperature. As shown in

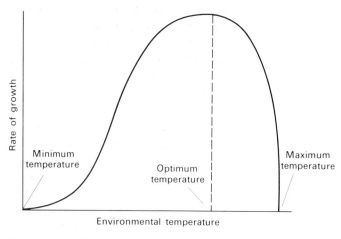

Fig. 9.1. The effect of temperature on the rate of growth of a micro-organism.

Fig. 9.1, there is a minimum temperature below which growth will not occur; as we rise above this the rate of growth increases in accordance with the laws governing the effect of temperature on the chemical reactions that make up growth. These are mostly enzyme catalysed. However, a point is reached when there is also a very rapid increase in the rate of inactivation of heat-sensitive cell components such as enzymes, ribosomes, DNA and membranes. Above an optimum temperature, this heat denaturation will occur so rapidly that there is a corresponding rapid drop in the rate of growth to give a maximum temperature for growth for that particular micro-organism.

Most micro-organisms are capable of growth in a temperature range of 20–30°C. It appears that the whole structure, metabolic machinery and control systems in a cell are such as to allow proper functioning only within such a range, although the actual values of the minimum, optimum and maximum growth temperatures vary considerably. Most micro-organisms have a growth optimum between 20 and 40°C and are called *mesophilic*. However, those inhabiting cold environments such as polar areas can grow at much lower temperatures; large numbers of micro-organisms are found on the surface of such unpromising areas as glaciers where they may even cause a visible green or red colour. These organisms are called *psychrophilic* and may cause trouble in food stored in refrigerators over a long period. Other micro-organisms called *thermophiles* are able to inhabit environments such as hot springs or compost heaps. These fascinating creatures can grow at temperatures as high as 80–100°C, when the great majority of living organisms would rapidly die. In fact, it seems that micro-organisms can grow at any temperature as long as water is in a liquid state. We do not completely understand how they do this, but the answer probably lies in a dual effect of an increased stability of many of the components of the cell coupled with more active repair mechanisms of heat-denatured components.

A distinction must be drawn between growth and survival. Although most micro-organisms are rapidly killed above their maximum growth temperature, they are not necessarily killed below their minimum growth temperature. Indeed, the ability of most bacteria to survive at liquid nitrogen temperatures is the basis of one of the most popular methods for preserving micro-organisms—a suspension is simply stored in liquid nitrogen.

pH

As with temperature, the pH range at which a micro-organism will grow varies considerably, although for a given species it usually covers about four pH units. Most organisms grow best at pH near neutrality, bacteria usually slightly on the alkaline side and algae and fungi on the acid side. However, some microbes can grow at extreme values of low or high environmental pH. For example, a few bacteria that obtain their energy by the oxidation of inorganic sulphur compounds to sulphuric acid can grow at a pH value as low as 0 (i.e. 1M H_2SO_4). Other bacteria causing infections of the human urinary tract can hydrolyse urea and this leads to the production of an excess of ammonia with a consequent rise of pH; these organisms grow at pH values as high as 11.

The inability of most bacteria to grow below a pH value of 3–4 is made use of in the food industry where pickling is a common method of preservation. Acetic acid in the form of vinegar may be added to the food or bacteria may themselves lower the pH value by fermentation. Another example of this process is in silage fermentation (see p. 155).

Oxygen and redox potential

The presence or absence of oxygen divides organisms into three main classes (Table 9.1). The basis of this differentiation lies mainly in the nature of the energy-producing systems and can be summarized as follows:

1 Some micro-organisms require oxygen as a terminal electron acceptor for oxidation (see p. 62) and if this is the only means of energy production, the organism will be a strict aerobe. A similar result will occur if the degradation of some

carbon and energy growth substrates such as hydrocarbon requires the entry of an oxygen atom or atoms into the molecule.

2 If, in addition, a micro-organism can obtain its energy in the absence of oxygen, it will be a facultative anaerobe. In this case, growth is usually more abundant aerobically than anaerobically.

3 Strict anaerobes have an energy-producing system which does not require oxygen, and in addition, they are actually poisoned by oxygen.

As would be expected, the aerobic or anaerobic nature of a micro-organism is related to the normal natural environment of that organism. Thus, methane-producing micro-organisms are strict anaerobes and live in an environment such as the lower gut of animals. On the other hand, methane-utilizing bacteria are strict aerobes and must occur in environments where oxygen is available from the air but also supplemented by methane as the result of the action of methane-producing bacteria in related anaerobic areas. As a result, in a stratified lake we find methane-producers in the anaerobic sediments, whilst methane utilizers are concentrated at the interface between the aerobic and anaerobic layers where methane is diffusing upwards and oxygen is diffusing downwards.

Table 9.1. The effect of oxygen on the growth of micro-organisms

	Growth in		
	Air	Low O_2	Absence of O_2
Strict aerobe	++++	+	−
Facultative anaerobe	++++ (or ++)	++	++
Strict anaerobe	−	−	++

Osmotic pressure

Most microbes are capable of growing within a fairly wide range of environmental osmotic pressure. Their ability to survive osmotic pressures lower than those of cytoplasm and thus to avoid lysis is related either to the presence of a mechanically tough cell wall or to a water-excreting mechanism such as the contractile vacuole. However, there are limits to the upper level of osmotic tolerance and the colonization of such environments as salt lakes, salt pans, and, to a lesser extent, the oceans requires specialized organisms called *osmophiles* or *halophiles*. Fortunately, most micro-organisms are unable to grow at such high osmotic pressures and this fact is made use of in the preservation of food by salting or by the addition of sugar.

Hydrostatic pressure

The only natural environments with hydrostatic pressures high enough to inhibit the growth of most micro-organisms are the depths of the oceans. Here the pressures may be as high as a thousand times those on the surface and we find colonization by specialized *barophiles*. Little is known about them since they will not grow at normal atmospheric pressures and special apparatus is required for their collection, isolation and study, as their exposure to atmospheric pressure causes them to explode.

Radiation

Most micro-organisms are killed by high doses of electromagnetic radiation, particularly in the ultraviolet range, and by smaller doses of ionizing radiation. Death occurs mainly by damage to DNA and variation in resistance usually reflects the differing abilities of cells to repair their radiation-damaged DNA. This is of importance to organisms suspended in the upper layers of the atmosphere when radiation may be a significant cause of death.

Visible light is essential to the growth of photosynthetic micro-organisms since it provides their energy source and these organisms usually have special mechanisms to avoid the deleterious effects of such radiation.

General conclusions

We can see that two overall groups of principles emerge:
1 A microbial species generally has a fairly wide range of environmental conditions in which it will grow. This range is usually higher in prokaryotes than eukaryotes as it is in micro-organisms compared with the cells of higher animals and plants. The width of this range can reflect either a less-sensitive cellular mechanism, or a capacity for controlling the composition of the cell in the presence of environmental extremes, in other words, a capacity for *homeostasis*. Both explanations can be considered. Thus, the ability to withstand a wide temperature range must be due to a difference in cell sensitivity while growth at different pH values may generally relate to the capability of regulating the pH of the cytoplasm so that it remains near neutrality.

2   Members of the microbial kingdom as a whole show an extraordinary ability to occupy extreme environmental niches. Yet again, this ability is particularly marked in pro-karyotes. For example, an extreme environment such as the Dead Sea or near the source of a hot spring will be populated by prokaryotes almost to the exclusion of eukaryotes. The advantages of the colonization of these extreme environments will be the lack of competition from other species for the available nutrients. Presumably the simpler structure of the prokaryotic cell is more capable of evolutionary adaptation or the genetic mechanisms are themselves more capable of change. On the other hand, the eukaryotic cell is more suited to the evolutionary change in the direction of a multi-cellular differentiated life form. There are therefore very few places on the earth's surface where physico-chemical conditions prevent microbial growth but there are other reasons for the widespread occurrence and importance of micro-organisms in natural environments. As a group they have tremendous metabolic versatility and are capable of degrading any naturally occurring organic material, and also the majority of synthetic compounds. They have the ability to sequester nutrients present at very low concentrations and this, along with their small size and potential short doubling times, means that they may be the most active group in many environments where their total mass is relatively small. In addition, bacterial spores are the most resistant biological structures known and actinomycete and fungal spores are the most highly evolved dispersal structures.

So we can envisage the environment as a vast depository of micro-organisms interacting with the environment, with each other and, as we saw in Chapter 8, with higher plants and animals. But before discussing the role of micro-organisms in geochemical cycles, let us first examine three specific environments, the atmosphere, aquatic ecosystems and the soil, from the point of view of microbial cells and populations.

The atmosphere

This is the simplest of the three in that it has a relatively uniform and constant composition, but the lack of nutrients, particularly water and organic carbon, mean that its major role for micro-organisms is as a medium for dispersal. Over land we typically find in the order of 100 organisms/litre of air near the ground, with fewer at higher altitudes because of difficulties of transport and the lethal effects of radiation. The majority are present as spores and most of these will be fungal spores, which are particularly well adapted to aerial dispersal through shape and structure and because of a wide range of spore release structures and mechanisms.

The microbiology of the atmosphere has three important applications for man. The first is in hospitals where such operations as bed-making and floor sweeping can create aerosols of pathogens, giving rise to variation in size and species composition of the airborne population. Spores released into air currents may be transported many miles before they settle through rainfall or impact. The second is in germ warfare where the aim is to maximize spore viability and infectivity, and to control when and where the cells come to ground. The third is the spread of plant pathogens, the majority of which are fungal and transmitted as spores. In all three cases it is important to know not only the mechanisms involved in the creation of such aerosols and the factors which cause them to settle, but also how long airborne organisms remain viable and infective.

Aquatic environments

In terms of volume, aquatic environments may be considered the major site for microbial growth, with 71% of the earth's surface covered by oceans, and growth believed possible at the greatest depths. Growth is generally slow, because of low temperature (90% of the world's sea water is always below 4°C) and low concentrations of organic matter. For example, the oceans contain 0.4−10 mg organic carbon/ml, and the concentration of freely suspended organisms is consequently low, in the order of 10−1000 organisms/ml. Organisms which can grow best at such low nutrient concentrations are said to be oligotrophic and have low saturation constants (K_s), or high substrate affinity (see p. 84). This gives them a competitive advantage over organisms with a high K_s even though the latter may have a higher maximum specific growth rate. Although suspended cell concentrations are low, cells accumulate in sediments, on the surfaces of plants, animals, and decaying organic material and on rocks and stones. Such surfaces afford protection and may also con-centrate nutrients, further to which bacteria produce extra-cellular polysaccharides and a range of appendages to facilitate attachment and colonization.

Of course, photosynthetic micro-organisms, the eukary-otic algae, blue-green bacteria and photosynthetic bacteria, require only carbon dioxide and light but thus require to be freely suspended near the surface of the water column. Many have evolved ways of adjusting their buoyancy, such as the

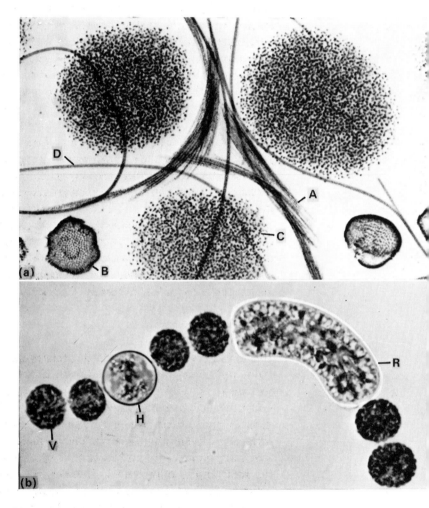

Fig. 9.1. (a) Planktonic blue-green bacteria from Esthwaite Water, English Lake District. A *Aphanizomenon*; B, *Gomphosphaeria*; C, *Microcystis*; D, *Oscillatoria* (×135). (b) Part of a thread of the blue-green bacterium, *Anabaena.* V, vegetative cell with gas vacuoles; H, heterocyst; R, young resting spore (akinette) (×1350). (Photographs kindly supplied by Hilda Canter-Lund.)

gas vesicles of cyanobacteria enabling vertical movement within the water column, and stratification of different groups of algae and blue-green bacteria occurs according to their preferences for different wavelengths of light. The growth of the algae in rivers or lakes is often limited by the level of minerals and, in particular, by nitrate, phosphate and sulphate. Various factors have led to a much increased rate of addition of such nutrients to inland waters. This process of enrichment is called *eutrophication* and it can be caused by various types of pollution such as the addition of treated sewage or of industrial waste or by drainage from agricultural land which is being increasingly intensively farmed. The result of this eutrophication is initially seen as the formation of rich algal blooms visible to the naked eye. The organisms responsible for such blooms are frequently planktonic blue-green bacteria (Plate 9.1), particularly those which possess gas vacuoles giving them sufficient buoyancy to stay in the well-lit surface layers. Unfortunately, such algal blooms can directly or indirectly lead to a marked deterioration in the quality of inland waters and can present serious economic problems by virtue of containing substances toxic to fish or of difficulties in filtration. Further, when the algae die, the decomposition of their cells by bacteria may lead to anaerobic conditions which, if extended through the body of water, may lead to the death of fish and other animals.

The soil

Soil represents the most varied and heterogeneous environment for micro-organisms, and in fact its structure and formation are dependent on microbial action, as well as physical and chemical processes. It has solid, liquid and gaseous phases; the first of these consisting of complexes formed between clay minerals and organic matter. The remaining volume, the pore space, usually constitutes 50% of the total and is filled with 'soil water' containing soluble organic and inorganic material, and finally the soil atmosphere which is water saturated, with less oxygen and more carbon dioxide than the atmosphere above the soil. We must therefore now consider the environment immediately surrounding the microbial cell, i.e. its microenvironment or microhabitat. Conditions in such microhabitats may be very different to those in the bulk soil. For instance, the cell may be growing on a piece of decaying plant material and may have an excess of carbon and nitrogen while the soil as a whole may be nutrient deficient. The existence of microhabitats means that processes requiring completely different conditions need not be spatially separated. For instance, aerobic organisms may grow on the outside of soil crumb or microbial film while anaerobic processes occur in the centre. This increases the scope for microbial interactions which might otherwise be reduced by relatively slow movement of materials and cells through soil. In addition, most microbial activity occurs on the surface of soil particles where nutrient availability and general environmental conditions, e.g. pH, are different to those of the soil as a whole.

The number of micro-organisms decreases with soil depth but in general one can estimate that 1 g of a typical garden soil will contain 10^7 bacteria and 5 metres of fungal mycelium. It has been estimated that 1 hectare of a typical agricultural field contains almost 6 000 kg wet weight of microbial biomass, which is the weight of 80 sheep. A major function for the microbial ecologist is to assess the significance and role of this biomass. Are the cells viable or dead, growing, active or dormant, vegetative cells or spores? Indeed, much soil activity comes from enzymes which have been secreted by cells or released following lysis and which then become complexed with clay minerals thus gaining protection from degradation and inactivation. Examples of such enzymes are urease, phosphatase and dehydrogenase and they are believed to be responsible for much of the decomposition of their substrates. They can remain active over long periods of time by binding to clay and humic colloids and their presence is therefore indicative of past, rather than current, microbial growth and activity.

Cycles of elements and matter

We have seen that the main elements from which living organisms are made up are carbon, hydrogen, oxygen, nitrogen, sulphur and phosphorus. Growth, considered over the whole range of living organisms on earth, consists of the conversion of these elements present in an inorganic form to the organic compounds that make up living matter. The energy for this conversion is ultimately derived from solar sources by photosynthesis.

$$\text{Inorganic forms of elements} \quad \xrightarrow[\text{Solar energy}]{\text{Living organisms}} \quad \text{Organic forms of elements}$$

If this were the only process occurring, life would soon cease as the inorganic forms of the elements and, in particular, of carbon and nitrogen were bound up into organic matter. In fact, the reverse process called mineralization must also occur and is brought about by the activity of living organisms so that cycles of elements and of matter occur.

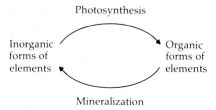

The situation is complicated by the existence in nature of oxidized and reduced states of most of the essential elements; organisms may only be able to use one or other form and further cycles therefore exist between them.

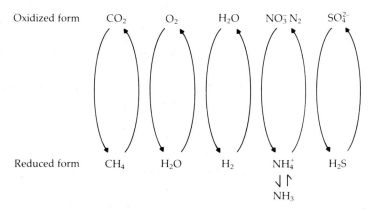

We must remember that micro-organisms interconvert these compounds for their own growth and each microbial process involves consumption of energy. While the different elements are cycled, energy flows through the ecosystem and is ultimately lost as heat. Let us consider in more detail the way in which carbon undergoes cyclic changes.

The carbon cycle

The major inorganic source of carbon for the growth of living organisms is carbon dioxide either in the atmosphere or in solution in surface waters. It is converted to an organic form by the action of autotrophic organisms which can use it as a sole source of carbon. The most important autotrophs in this process are those carrying out an oxygen-producing type of photosynthesis—the seed plants on land and the algae in water. Photosynthetic and chemosynthetic autotrophic bacteria also play a role, though a relatively minor one.

$$CO_2 \xrightarrow[\text{Autotrophs}]{} \text{Organic carbon}$$

A small part of the inorganic carbon will be in the reduced form of methane which can be utilized as the sole source of carbon and energy by a special group of aerobic bacteria which convert it into organic carbon and CO_2.

$$CH_4 \xrightarrow[\substack{\text{Methane-utilizing} \\ \text{bacteria}}]{\nearrow \quad CO_2} \text{Organic carbon}$$

The rate of conversion of inorganic to organic carbon is such that in 20–30 years the total CO_2 in the atmosphere would be completely exhausted in the absence of replenishment from the oceans. However, the reverse process of the mineralization of organic to inorganic carbon by the action of heterotrophs prevents this exhaustion; the major end-product is CO_2 but some bacteria will produce CH_4 by anaerobic respiration and fermentation.

$$\text{Organic carbon} \xrightarrow[\text{Heterotrophs}]{(CH_4)} CO_2$$

Thus, a carbon cycle is built up to give an equilibrium with equal rates in both directions:

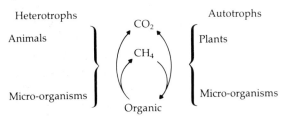

Although the conversion of inorganic to organic carbon by the plant kingdom and by autotrophic micro-organisms is relatively straightforward, the reverse process by the animal kingdom and heterotrophic micro-organisms is more complex and must be considered in greater detail.

The mineralization of organic carbon

Heterotrophic organisms, be they animals or microbes, obtain their carbon and energy by the metabolism of an organic source provided by another form of life.

$$\text{Organic carbon} \xrightarrow[CO_2(CH_4)]{\text{Heterotroph}} \text{Organic carbon}$$ (Cells and organic end produc

As a result, some of the organic carbon is mineralized while the rest is converted to further organic carbon in the

154

form of new growth of the heterotroph or as end-products of metabolism. The efficiency of this conversion will vary according to the organism and to the aerobic or anaerobic nature of the process. The particular heterotroph and any organic carbon it may produce then act as the food source for other heterotrophs and so on along a food chain. At each stage of this food chain a percentage of the original organic carbon is mineralized until the process is complete.

$$\text{Autotroph} \xrightarrow[\text{Heterotroph}]{\text{CO}_2(\text{CH}_4)} \text{Organic carbon} \xrightarrow[\text{Heterotroph}]{\text{CO}_2(\text{CH}_4)} \left(\begin{array}{c} \text{CO}_2(\text{CH}_4) \\ \text{Organic carbon} \\ \text{Heterotroph} \end{array} \right.$$

Micro-organisms play an exceedingly important part in this process. Any compound which is a component of a living organism must be susceptible to mineralization or it would eventually accumulate on the surface of the earth so that ultimately all the carbon would be in this unavailable form. We have seen that micro-organisms as a whole have a very wide ability to break down organic compounds and they can in fact metabolize all naturally-occurring chemicals. It is a matter of common observation that dead animal or plant material falling on soil or in water is fairly rapidly destroyed. This occurs by the sequential action of a variety of micro-organisms, each utilizing one or more component compounds. If any substance accumulates, then the growth of a microbe capable of utilizing it is favoured. However, some plant components in soil are used only slowly and some of them, together with the products of saprophytic microbes, constitute the humus, a complex mixture of materials which helps to maintain a suitable texture for plant growth.

An important factor affecting the rate and extent of mineralization is the availability of oxygen. If conditions become anaerobic, as for example in a waterlogged soil, then the accumulation of acids and other inhibitory end-products of fermentation prevents further microbial growth and metabolism. In this way, layers of partly decomposed plant material can accumulate as in peat deposits. Such a phenomenon of partial attack is made use of in silage production. Suitable plant material, such as grasses, is packed into cylindrical towers or pits called silos, where the conditions rapidly become anaerobic. Carbohydrate fermentation by bacteria leads to the accumulation of lactic acid in the silo. The pH falls to a level which prevents the growth and activity of organisms with the potential to cause deterioration and quickly a stable situation is reached in which the silage can be stored for long periods prior to its use as animal fodder.

The action of micro-organisms, then, is of fundamental importance in maintaining the carbon cycle and all naturally-occurring organic compounds are eventually mineralized to CO_2 and CH_4. However, man is introducing increasingly large amounts of synthetic organic compounds into the environment, particularly as herbicides and pesticides. Many are not readily subject to microbial attack and will therefore accumulate. In most cases we do not know what will be the effect of this accumulation and apart from any direct toxicity to man we cannot usually assess how far normal food chains will be broken with equally disastrous results. Clearly, if such chemicals must be used, care should be taken to ensure that they are non-toxic to man, but they should also be designed to be eventually subject to microbial mineralization—in other words, they should be biodegradable, if only slowly. An example of what can be done can be seen in the case of synthetic detergents. In the early days of their commercial production, such detergents were not biodegradable and caused considerable problems as a result of their accumulation, particularly in streams and rivers where layers of foam built up. Legislation led to the design of detergents which are subject to microbial attack and the problem was apparently solved. However, a further problem has arisen. Most detergents contain phosphate; microbial attack liberates inorganic phosphate which may itself cause pollution problems by eutrophication (see p. 151); the next stage must be to design biodegradable compounds which do not contain phosphate.

An interesting question arising from the contamination of our environment with both natural and man-made chemicals is how far can mutation of micro-organisms effect changes in the substrate-specificity of enzymes so as to allow quite different and new compounds to be attacked? In other words, how far can pollution be overcome by the gradual evolution of organisms with new metabolic capabilities? Certainly, we can show this process in the laboratory provided that the new substrate is sufficiently close in structure to a normal substrate. However, the time required for the evolution of microbes attacking compounds like polyvinyl chloride or nylon may be considerable, if it ever occurs.

Sewage treatment. The development of big cities has resulted in the production of large amounts of organic wastes in a restricted area. If the untreated sewage is simply discharged into nearby waters, as is all too common, two problems may result. Firstly, there is a public health hazard caused by contamination with potentially pathogenic micro-organisms. Secondly, the affected waters may be made anaerobic by the action of microbial aerobic metabolism on the dissolved organic compounds; animal life is destroyed by the resultant lack of oxygen. Consequently, some treatment of raw sewage is desirable to decrease the level of organic compounds. Since this is essentially a process of mineralization, the ability of micro-organisms to carry it out is used in the design of sewage-treatment plants. There are two basic stages involved in such plants:

(a) *Sedimentation and anaerobic decomposition.* The solid particles in raw sewage are allowed to sediment in large settling tanks. The precipitate undergoes an anaerobic decomposition with the production of CO_2 and CH_4 and the undigested material is removed periodically from the bottom of the tank.

(b) *Aerobic decomposition of soluble fraction.* The soluble matter in sewage is subject to aerobic microbial decomposition. This may be brought about by a trickling filter in which the liquid sewage is sprayed onto the surface of a bed of crushed stone impregnated with mineralizing microbes. Another method is based on the fact that if air is passed through liquid sewage, a precipitate is formed which becomes rich in micro-organisms. This precipitate or 'activated sludge' as it is charmingly called, can be added to further aerated soluble sewage and will cause a rapid mineralization. The processes are represented in Fig. 9.2.

The nitrogen cycle

Nitrogen is present in nature in organic compounds, e.g. proteins, and in many inorganic forms, the most reduced being NH_3 and the most oxidized being NO_3^-. However, a suitable source of nitrogen is often the factor limiting growth in an ecosystem. Although there is a plentiful supply of molecular nitrogen in the atmosphere, we have seen that there is a restricted group of nitrogen-fixing micro-organisms capable of using it (Chapter 4) and nitrogen fixation requires high energy input. Prokaryotes alone are capable of biological nitrogen fixation and the most important group agronomically are rhizobia, which form root nodule associations with leguminous plants (see p. 128). There is a degree of specificity between the infecting bacterial strain and the host plant and cross-fertilization groups have been established. Knowledge of such groups allows us to increase nodulation of specific crops by inoculation of seeds with the appropriate *Rhizobium* strains. Other associations are also of agricultural importance, in particular the formation of root nodules on

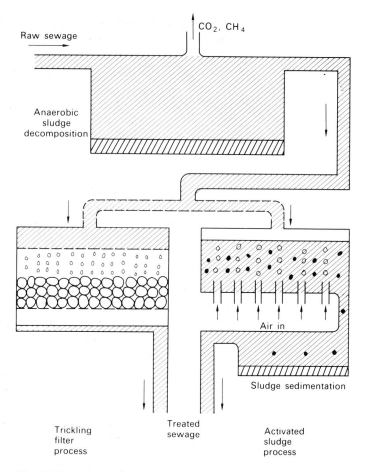

Fig. 9.2. Two types of sewage treatment plant.

non-leguminous plants by actinomycetes (*Frankia*) and fixation by blue-green bacteria in rice paddy fields.

Nitrogen fixation represents the rate-limiting step within the nitrogen cycle and it is, therefore, imperative that there should be efficient cycling between organic and inorganic forms. Inorganic nitrogen in the form of ammonia or nitrate is converted to an organic state mainly by the action of plants and micro-organisms (if nitrate is used it must be first reduced to ammonia). We have already discussed the differences between micro-organisms in the ability to synthesize their amino acids, purines, pyrimidines and other organic nitrogenous compounds for themselves and therefore in their ability to use an inorganic source of nitrogen (Chapter 4). The responsibility for the mineralization of this organic nitrogen to ammonia lies largely with micro-organisms either by their action on dead organisms or by their degradation of animal excretory products.

In traditional farming systems, mineralization has provided sufficient ammonia for plant growth but cannot satisfy the nitrogen requirements of modern, intensive agriculture. This has led to a dramatic increase in chemical fixation of nitrogen by the Haber process such that 25% of all fixation is now carried out industrially for production of ammonium-based fertilizers. Their application can, however, lead to environmental problems. Ammonia is converted by nitrifying bacteria first to nitrite, which rarely accumulates, and then to nitrate. While ammonium ions are bound to negatively charged clay particles in the soil, nitrate ions are readily leached and can reach high concentrations in run-off waters and rivers. This can cause methaemaglobinemia, or 'blue-baby disease' in animals and infants and can lead to eutrophication. As well as increasing nitrate pollution, nitrification results in loss of nitrogen from the environment either through leaching or through subsequent anaerobic reduction of nitrate, through denitrification, to gaseous oxides of nitrogen and nitrogen gas. Nitrification is therefore often actively discouraged by addition of inhibitors, e.g. N-Serve, along with ammonium-based fertilizers.

The nitrogen cycle is summarized in Fig. 9.3 but it is worth noting here that none of the nutrient cycles occurs in isolation. For example, decomposition of organic compounds will simultaneously release carbon, nitrogen and other elements and each micro-organism is involved in the cycling of many nutrients.

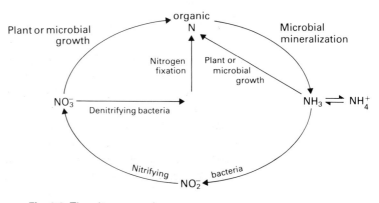

Fig. 9.3. The nitrogen cycle.

The sulphur cycle

Sulphur, like nitrogen, is found in organic compounds and a variety of inorganic forms and it too may be deficient in soils. Sulphate is the major form assimilated by plants and micro-organisms, and it is reduced intracellularly before incorporation into proteins and other organic compounds.

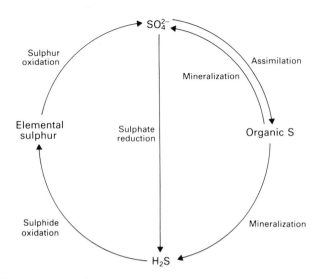

Fig. 9.4. The sulphur cycle.

Mineralization of these compounds results in production of sulphate or, under anaerobic conditions, hydrogen sulphide.

Hydrogen sulphide may also be produced anaerobically by dissimilatory reduction of sulphate. This process is carried out by sulphate reducing bacteria, e.g. *Desulphovibrio, Desulphomonas* and *Desulphomaculum*, which use sulphate as an alternative electron acceptor to oxygen. Again, this process illustrates the links between the different nutrient cycles as sulphate reduction accounts for up to 50% of the carbon mineralized in marine sediments. The process gives rise to the black odorous nature of muds and sediments of estuaries, the colour arising through precipitation of ferrous sulphide. It may also be of significant economic importance, as sulphide is believed to have a role in corrosion of mild steel.

Hydrogen sulphide is oxidized chemically to sulphate in the presence of oxygen but can be oxidized under aerobic and anaerobic conditions by micro-organisms. Anaerobic conversion is carried out by the photosynthetic purple sulphur and green-sulphur bacteria (e.g. *Chromatium* and *Chlorobium*). Aerobic oxidation of sulphide, and of other inorganic forms such as thiosulphate, tetrathionate and elemental sulphur, is carried out by sulphur oxidizing bacteria, e.g. *Beggiatoa* and *Thiobacillus*. They too can be of economic significance and sulphuric acid produced by *Thiobacillus* is responsible for the development of cracks in concrete cooling towers.

10 : Micro-organisms in Industry

In theory, the formation of any microbial metabolite can be made the basis of a commercial process. However, apart from the industrial need for the product, the most important question must always be the relative cost of a process based on the use of micro-organisms compared with a chemical synthesis or an alternative biological source. In the past, most systems have been concerned with the production of relatively simple extracellular end-products or secondary products of metabolism which usually occur in large quantities. However, the recent advances in genetic engineering have broadened the potential of microbially-based processes. Further, as our understanding of the physiology of micro-organisms has grown, and as the consumer demand for a wide range of products ranging from soft drinks to sophisticated pharmaceuticals has increased, the industrial potential of micro-organisms has come to be recognized.

It is possible to distinguish a series of phases in the development of industrial microbiology (Table 10.1). The first phase is relatively ancient and involves all types of fermentation to produce alcoholic beverages and the traditional production of dairy products and bread. These processes share two characteristics, namely, they are non-aseptic processes and rely either on a large inoculum or on good housekeeping to prevent microbial spoilage of the product. The second common feature is that they all reflect the synthesis by the organism of a primary product of metabolism, e.g. the enzyme rennin in cheese making, lactic acid in yoghurt, carbon dioxide in bread, and alcohol in wines and beers. Behind this simple statement, it must be realized that although each of these is the major product of the microbial process, in addition the cells produce a wide variety of small

Table 10.1. Phases in industrial microbiology

> 4000 years ago	Non-aseptic growth
	Brewing, bread, cheese
100 years ago	Large scale, partial aseptic growth
	Yeast, metabolic end products
40 years ago	Aerobic, aseptic growth. Strain selection
	Antibiotics, amino acids
20 years ago	Continuous culture
	Single-cell protein
10 years ago	Biotechnology
	Enzymes, cloned-gene products

molecular weight by-products which play an essential role in the development of flavour.

The second phase of the exploitation of micro-organisms in industry was the development of large scale relatively sterile culture facilities in which an inoculum is added. This originated at the turn of the century with the large scale cultivation of yeast cells for the brewing and baking industries. Obviously, any change in the scale of a process increases the potential loss if the culture becomes contaminated with an undesirable organism. Hence there was and continues to be a need for a means of sterilizing large quantities of media, fermentation vessels and pipelines; steam was often used for this purpose. The growth media are often the by-product of other industries, e.g. complex mixtures of molasses, and yeast extracts (distiller's by-product), with the addition of some pure chemicals to provide the nutrient balance. Although contamination may occur in these processes, the use of a large inoculum minimizes the problem.

Most of the products in this second phase come from anaerobic microbial fermentations, e.g. acetone and butanol, and hence the industry that grew up was known as the *fermentation industry*. Unfortunately, the term has tended to be applied to other more recent developments where growth is essentially aerobic, e.g. the production of antibiotics by bacteria and fungi.

With the exception of alcoholic drinks and bread making, the fermentation industry was in danger of extinction when the discovery of the therapeutic use of antibiotics in the 1940s led to a rebirth of interest in its applications. Since most antibiotics are very complex chemicals, microbial production is usually much cheaper than chemical synthesis. Batch culture was used, but under aerobic conditions the need for aseptic culture under carefully controlled conditions had to be recognized from the outset. Research led to the selection of more efficient antibiotic producing strains which are jealously guarded by the responsible industry. For example, strains of *Penicillium* in current use produce as much as ten thousand times as much penicillin per unit volume of fermentation liquor than does Fleming's original isolate.

During the 1960s, interest in the use of micro-organisms as bulk sources of protein for animal and human food, the so-called 'single-cell protein', led to the exploitation of continuous culture methods on an industrial scale. This development necessitated the use of sophisticated fermentation equipment and, since the continuous fermenters can be run for long periods of time, the maintenance of complete asepsis became of paramount importance.

Finally, the most recent phase has been termed *biotechnology* and has gripped the imagination of both layman and scientist alike. Biotechnology is a term that has often been used indiscriminately and fashionably, but it can be defined as the application of biological organisms, systems or processes to the manufacturing and service industries. As such, it can be said to include most of the areas mentioned previously, although there has been some tendency to equate gene-cloning and recombinant DNA technology with biotechnology. These processes have been fundamental to a realization of the potential that exists, but they are only one aspect of this rapidly emerging phase of biologically-based industries. Of comparable importance is the use of enzymes and immobilized cells as catalysts and sensors, the manipulation of cell physiology for the maximum production of metabolites and the use of cell fusion to generate either monoclonal antibodies or new cell lines. It should also be stressed that although biotechnology is still primarily a microbiological industry, plant and animal cells are becoming increasingly important, their use involving similar methods to those pioneered by microbiologists.

Present day usage of micro-organisms in industry reflects all these phases of development. Thus, there are still the traditional breweries, and antibiotic and yeast production remain major industries running alongside the newer developments grouped together under the heading of biotechnology.

The ethanol fermentation

Historically, the most important microbial product has been ethanol, formed by yeast as a result of carbohydrate fermentation (p. 63). The simplest methods of production involve the use of a plant material with a high sugar content such as grapes, these forming a substrate to be fermented directly to produce wine; the type of wine will depend on the grape and on the variety of yeast. The process consists of crushing the grapes to a clear grape juice, *must*, which is fermented either by yeasts naturally present on the grapes or by the addition of pure *starter cultures* of appropriate wine yeasts. When the natural yeast flora present on the grape surface is used as the inoculum, then it is frequently the case that a microbial succession of yeast types takes place, with the dominant species being selected by the environmental conditions, e.g. sugar and alcohol concentrations, pH, etc. The major fermentation product is ethanol, but in terms of taste many of the minor by-products are more important. The fermentation is almost entirely due to yeast since the growth of contaminating bacteria is prevented by the high acidity and sugar content of the grape juice. After fermentation is complete or has reached a required stage, the yeast is removed and the wine is allowed to age by a non-microbiological process. Care must be taken to prevent microbial spoilage, particularly by the action of bacteria oxidizing ethanol to acetic acid which will occur if the conditions become too aerobic in the ageing process; if such an oxidation occurs, vinegar is the ultimate product. In some wines, a secondary fermentation is desirable. During sherry production, for example, ethanol conversion to acetaldehyde is deliberately encouraged, resulting in the characteristic change in the flavour of the wine. The yeasts involved in this process, *Saccharomyces ovarum* and *S. bayanus*, form a surface skin on the wine as it lies in a partially filled barrel. This is termed the *flor* and wines forming this layer give rise to fino sherry. The formation of petillant wines (those emitting a fine stream of bubbles) and champagne also require secondary fermentations by bacteria and yeasts respectively.

In beer production, the carbohydrate fermented is the starch present in barley, corn or rice. Since alcohol-producing yeasts are unable to break down starch themselves, a preliminary depolymerization stage is required. For this purpose, use can be made either of the starch-hydrolysing amylases present in barley as occurs in European beer production, or of fungal amylases which are added in Japanese sake production. Following this initial depolymerization, special brewer's yeasts produce the ethanol. Different types of beer are made by using different yeasts, by the addition of hops or by the use of a slower, low-temperature fermentation in lager production. As with wine, the flavour of beer is very much related to the metabolism of the yeast. However, whereas with wine the secondary products arise primarily from sugar, in beer the flavour compounds are derived from the amino acids released from barley proteins degraded during malting.

In Europe, beer is produced using the following general procedure. Germination of the barley seeds is encouraged, leading to the production of amylases and proteases which liberate sugars and amino acids from the stored starch and proteins. Before the seeds can begin to grow, germination is stopped by gentle heating which leaves the enzymes intact and functional. The germinated barley is then ground up and mixed with hot water to allow enzyme activity to proceed, and all of the starch and protein to be broken down to mixtures of sugars and amino acids. This mix is called the *wort* and, if bitter is to be made, hops are added. A yeast starter culture is added and the organism will start to grow. The process is initially aerobic as yeast needs oxygen in order to make its cell membrane sterols. As a result of its fermentation, yeast produces one molecule of carbon dioxide per molecule of alcohol. This gas is more dense than air and forms a frothy layer on top of the fermentation excluding the entry of further air. Once the oxygen runs out, growth stops and fermentation begins in earnest. The lack of oxygen has many consequences for the physiology of the yeast which are an inherent part of the flavour formation in beer production. Another important consequence is that the layer of froth acts as a barrier to the contamination of the fermentation by bacteria which might cause spoilage. For this reason, beer manufacture, until recent times, was carried out in large open vats.

There are two further uses of yeast alcoholic fermentation. One is industrial ethanol production by distillation, although the more competitive chemical synthesis of ethanol has largely displaced microbial fermentation, that is in all except those countries where oil is expensive and fuel for driving the distillation process is still cheap. The other process is in breadmaking where the leavening stage is due to a yeast fermentation producing carbon dioxide.

The ability of micro-organisms to synthesize a wide variety of fermentation end-products has been made the basis of what was a highly developed fermentation industry. Substances such as glycerol, lactic acid, butylene glycol, acetone, butanol, acetic acid and butyric acid have been made this way. However, in the last thirty years, rises in the cost of suitable carbohydrates as raw materials and the development of more efficient petrochemical syntheses has made the microbial methods increasingly uneconomic. As the balance of the economics again swings in favour of microbially-based processes and our knowledge of microbial physiology and regulatory mechanisms becomes more complete, this area is again attracting attention as a potential 'new' industry.

The production of antibiotics and single-cell protein

Two processes mentioned previously will now be considered in more detail:

1 Antibiotic production as a group of relatively high value extracellular secondary metabolites.

2 Single-cell protein as a very large scale low value whole cell product.

In both cases, the development of a commercially viable process involved the interaction of scientists from many different disciplines. Microbiologists, geneticists, biochemists and engineers were required both to overcome their prejudices about each other and also to solve the problems of large scale microbial cultivation under carefully controlled conditions. There was the need for the design of aerated fermenters, sterilization plants, product extraction regimes, as well as the selection of the ideal organisms by comparison of naturally-occurring strains and their subsequent mutagenesis to produce better strains. Just as important was a need to carry out process design. It is necessary to decide on the design of the growth medium, the rate of addition of various compounds (inducers, etc.; see below), the shape of the fermenter, whether batch or continuous culture is most suitable, and stages in the work-up of the organisms; all of these will contribute to the productivity and hence commercial viability of the process. Let us examine some of these considerations.

(a) *Design of the growth medium.* This may seem trivial at first sight, but it is often crucially important. For example, antibiotics are usually only synthesized at the end of the exponential phase of growth, i.e. they are secondary metabolites. To get a good fermentation, one needs a high concentration of biomass and hence a continuing supply of carbon and nitrogen sources and of oxygen. However, as the microbial biomass increases, the ease of oxygenating the culture is reduced and so a high biomass concentration is a mixed blessing for the fermentation engineer! Thus, the nutrient balance of the medium is controlled to regulate growth and in the subsequent production phase it is optimized for penicillin production (Fig. 10.1).

With a high value product such as an antibiotic, the cost of nutrients is not a major factor in the cost of the final product. However, in the case of single-cell protein, the choice of a cheap and easily available carbon and energy source for growth is all important in determining the cost of

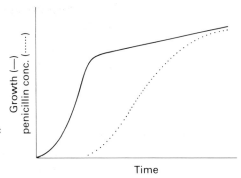

Fig. 10.1. Typical growth curve for antibiotic production. Note the two phases of growth, an initial rapid exponential growth phase followed by slower production phase.

the product. The main substances considered have been either fossil fuels or their derivatives (n-alkanes from oil, methane from natural gas or methanol by chemical oxidation of methane) or carbohydrates (especially starch). The cost of these substrates relative to the cost of alternative conventional sources of protein such as soybean meal and fish meal will determine the commercial viability of the process.

(b)   *Medium additions, 'steered fermentations'.* Early in the development of production, it was observed that the chemical identity of the penicillin produced by American scientists was different to that of their British counterparts. The effect was traced to a high concentration of phenylalanine in the American growth medium which was deaminated by the organism to phenylacetic acid which can be incorporated into the penicillin molecule to give benzylpenicillin. It was found to be desirable to produce this variant molecule and so phenylacetic acid is now added to the medium to 'steer' the fermentation towards the production of benzylpenicillin. There are now many examples of such additions both in the antibiotic industry and in the production of other complex molecules by micro-organisms. The basic principle of a steered fermentation is to supply a part of the molecule which the cell has difficulty in synthesizing or which it would not normally synthesize under production conditions, in order to steer the production pathway towards a particular end-product.

(c)   *Design of the fermenter.* Basically, the fermenter is a large tank with a device for mixing the contents. In reality, the design of the vessel and the provision of a means of mixing the contents of the fermenter and the related problem of aeration are very important factors in maintaining sterile conditions and productivity. Both of our processes depend upon aerobic

167

growth and the supply of oxygen can be a limiting factor in determining their viability. There are three main methods of aeration in industry (Fig. 10.2).

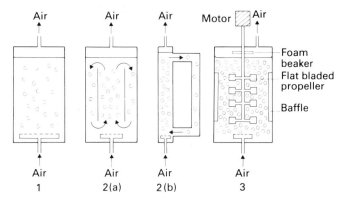

Fig. 10.2. Designs for aeration and stirring in fermenters. 1) The aerated tower. 2) The airlift fermenter, a) simple design; (b) ICI pressure-cycle fermenter. 3) Stirred fermenter.

1 *The aerated tower.* Air is simply sparged into the base of the vessel; this gives cheap, but relatively poor, aeration.

2 *The airlift fermenter.* Air is sparged through only part of the fermenter to cause an efficient circulation within the whole vessel. An example is in the ICI single-cell protein plant, which is the most radical design to emerge in recent years. *Methylococcus methylotrophus* is grown on methanol as the sole carbon and energy source to give a product called Pruteen, which is used as an animal feed.

3 *The stirred fermenter.* Here, aeration depends upon a mechanical stirrer to give the most efficient, but also the most costly, aeration. Although the method is commonly used in the laboratory, on an industrial scale the cost of supplying power to the stirrer can become prohibitively expensive.

It should be noted that the method used for stirring also mixes the fermenter contents and may be important for cooling. As larger and more productive fermenters are designed, the surface area to volume ratio decreases and hence the heat produced as a by-product of microbial growth cannot be so easily dissipated. Instead of the need for the supply of heat in a laboratory-scale fermenter, a cooling system has to be designed.

A final, but key question concerns the choice of batch or continuous culture. For many new processes, the greater productivity, the ease of optimizing and maintaining environmental conditions and the overall efficient use of the industrial plant in continuous culture makes it the method of choice, as in the single-cell protein systems. However, batch cultures will probably continue to be used for secondary metabolites such as penicillin, for traditional processes and for small-scale systems.

(d) *Scale-up of the organism*. Many of the organisms used in industry are bred for high productivity, but as a result they are often unstable and when revertants arise they often have a faster growth rate than the production organism. Thus, the scale-up of the culture from the original inoculum (often just a few thousand fungal spores in an antibiotic fermentation) to a 50 000 l fermenter requires great care. Generally, this is achieved in several consecutive stages of increasing volume leading to an inoculum suitable for the production fermenter. At each stage, the potential productivity of the inoculum is checked and only the best inocula are used. In continuous culture, the chances of reversion of the culture become even greater with the long time scale of the fermentation, and selective processes may have to be built into the system to ensure stability.

Recombinant DNA technology and industrial microbiology

The commercial production of antibiotics and of single-cell protein acted as stimuli to the microbial fermentation industry. Conventional genetics made it possible to 'tailor' micro-organisms for the production of almost any microbial metabolite be it vitamin, amino-acid or enzyme. Indeed, antibiotic production is an illustration of what can be done, albeit in a rather empirical way. The genetics of micro-organisms allow considerable sophistication in the methods used. However, the limitation has always been the information inherent in the organism's genome or at very best in that present in related species with which genetic exchange was possible. Gene cloning extends the genome of the micro-organism by allowing the introduction on novel genes from comparatively unrelated species. Thus, as was discussed earlier, the application of protoplast fusion to Actinomycetes may result in new biosynthetic pathways and possibly new antibiotics. The cloning of genes from higher eukaryotes, in particular from man and his associated domestic animals, has been seen to offer even greater industrial potential. What micro-organisms should we then use as universal recipients for such genes and hence as production organisms? The two most commonly in contention by virtue of our understanding of their molecular genetics are the prokaryote *Escherichia coli* and the eukaryote *Saccharomyces cerevisiae*. Some of the important products which gene cloning may make available in the future are listed in Table 10.2.

Table 10.2. Proteins likely to be produced by gene cloning

Protein	Use
Insulin	Control of diabetes
Growth hormone	Control of dwarfism and growth promotion in cattle
Interferon	Antiviral agent and potential anticancer drug
Ricin	Anti-cancer agent
Viral antigens	Vaccines
Blood proteins	Control of blood clotting

Given that micro-organisms grow rapidly and economically under environmental conditions which can be carefully optimized, large quantities of hormones (e.g. insulin and growth hormones) and natural pharmaceuticals (e.g. the interferons) could be produced by 'fermentation' very much more cheaply and on a larger scale compared with conventional methods. Thus, human growth hormone was previously extracted from the pituitary gland of cadavers and was frequently in short supply for the treatment of growth defects. An increase in the supply not only ensures that all patients can be treated cheaply, but allows the valuable research which may result in new therapeutic uses for the protein. Equally important has been the use of gene-cloning for the development of new vaccines which could not readily be made by conventional routes due to the virulence of the virus or pathogenic organism. Now genes for single antigens can be cloned and expressed by bacteria and a purified antigen which has not been derived directly from the pathogenic organism or virus may be used as a vaccine. In this way, vaccines for viral hepatitis and foot and mouth disease have been developed.

Microbial cells and enzymes

As was stressed above, the new wave of microbial biotechnology does not simply involve gene cloning. In the last few years, several commercial processes using immobilized microbial cells and enzymes have been developed with important economic consequences.

Enzymes have been used in industry for over 70 years, initially in detergents, which is still probably one of the largest

170

bulk uses of proteases and lipases. There has also been interest in the use of microbial enzymes in conventional biological industries, e.g. rennet for cheese manufacture and proteases and amylases for use in malting. However, the last ten years have seen the development of immobilized enzymes and cells as production systems. This change stems from several advantages of enzymes over chemical processes:

1 Enzymes carry out stereospecific reactions with great accuracy, whereas the appropriate chemical technology results in many often useless side-products from which the desired product must be purified.

2 Enzymes are cheap and carry out reactions at low temperature and at atmospheric pressure. Chemical catalysts on the other hand are often expensive and frequently need very special conditions for their reactivity.

3 There is great diversity in the metabolism and habitat among micro-organisms and thus it is possible to find an enzyme which carries out a desired reaction under appropriate conditions merely by screening a wide range of organisms.

4 Even when the enzyme does not have precisely the right properties, well-established genetic procedures can be used to isolate mutants with altered enzyme function. Thus, enzymes can be isolated having different substrate specificities or with different physical properties, e.g. temperature resistance. Even when this approach has been exhausted, molecular biologists are currently developing techniques for making very specific changes in genes. Both conventional genetics and gene-cloning can be used to increase the expression of the desired enzyme.

These advantages give enzymes considerable potential as replacements for chemical processes. More recently, whole microbial cells have been used for specific chemical transformations. This should not be confused with an anaerobic fermentation or conventional secondary metabolite production. Only part of the cell's metabolism is now being utilized, usually a single pathway, and sometimes only a single enzyme. The advantage of using cells is that the expense of purifying the enzyme is avoided and, in some cases, the enzyme is more stable in its natural environment than after purification. Frequently, enzymes and cells are subjected to immobilization on an inert support (Fig. 10.3). The rationale for immobilizing enzymes is that generally their stability is improved and there may be subtle alterations in their properties.

Immobilization affords a simple way of separating the enzyme or cell from the products when the reaction is com-

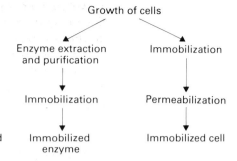

Fig. 10.3. Preparation of immobilized enzymes and cells.

plete and is likely to prove invaluable in the development of biological sensors (biosensors)—special electrodes based upon the selectivity and high affinity of enzymes for their substrates. The enzyme reaction, which can measure tiny quantities of a chemical in a complex mixture, results in an electrical signal passing to an appropriate measuring device. In this way, a continuous and very accurate monitoring system can be established for any compound. Already such devices, based on microbial enzymes, are helping diabetics balance their sugar intake.

Let us consider some examples of the use of isolated and immobilized enzymes in a little more detail.

(a) *Sweeteners*. Sucrose has been used for a long time as a sweetener in many foods and soft drinks. Certain bacteria and fungi produce an enzyme, glucose isomerase, which converts glucose to an equimolar mixture of glucose and fructose, which has been found to be sweeter than the equivalent sucrose solution. Glucose itself is not particuarly sweet, but unlike sucrose it is readily and cheaply available in large quantities from the breakdown of starch by amylases.

$$\text{Starch} \xrightarrow{\text{Amylases}} \text{Glucose} \xrightarrow[\substack{\text{Glucose} \\ \text{isomerase}}]{} \text{Hig fructose syrup}$$

Consequently, the production of high fructose syrups from starch has been seen to be a way of replacing an expensive food ingredient (sucrose) with a cheaper one (high fructose syrup). In the last few years, these syrups have all but eliminated the use of sucrose in confectionary and soft drinks. The production of high fructose syrups exemplifies the range of forms in which microbial enzymes are used. In some cases, the enzyme is subjected to a degree of purification to remove other undesirable enzymes, while in other processes the whole micro-organism is immobilized by an agent which forms cross-links between the cells. The process may involve

172

the enzyme being packed into a column and passage of the glucose syrup through the column. In other systems, the enzyme is simply incubated with the glucose syrup in a stirred tank and no attempt is made to recover the enzyme from the product. In this case, the method of stabilizing the enzyme must itself be non-toxic since it will pass into products for human consumption.

(b) *Amino acids.* Non-fattening sweeteners have recently been developed as diet aids and to counter obesity and associated illnesses. Although saccharine has been available for many years, concern about possible health hazards and an unpleasant after-taste in certain formulations has created a market for alternative products. Some of these are proteins of plant origin which will almost certainly eventually be produced via gene-cloning, e.g. thaumatin. However, one compound exciting great interest is called 'Aspartame' and is a modified dipeptide of aspartic acid and phenylalanine. The size of the market for phenylalanine illustrates both the potential of micro-organisms in production processes and the impact of new products on the demand for biological molecules. Thus, the world market for phenylalanine, previously mainly used for therapeutic purposes, increased forty-fold between 1981–1984 and is still increasing.

Amino acids are also used in the pharmaceutical industry and as feed supplements. They are produced by two principal routes: fermentation and immobilized cells and enzymes. The major difference in these two approaches is that the fermentation route starts from sugars and involves many enzymic conversions. When enzymes are used, then usually only one or a few transformation steps are involved and cells are treated chemically or physically to inhibit enzyme activities which might interfere with the desired chemical conversion.

(c) *Antibiotics and pharmaceuticals.* Here the best examples are the modification of penicillin to change the structure of the molecule and the hydroxylation of steroids to generate more potent drugs. Penicillin acylase catalyses the deacylation of the sidechain of the major penicillin produced by industry to yield 6-aminopenicillanic acid (6-APA). Semi-synthetic penicillin molecules can then be made by chemical routes from the 6-APA, e.g. ampicillin from phenylglycine (Fig. 10.4).

Penicillin acylase is produced by *Escherichia coli* and both immobilized cells and the purified enzyme have been investigated as the basis of a commercial process. Current processes use the immobilized enzyme.

173

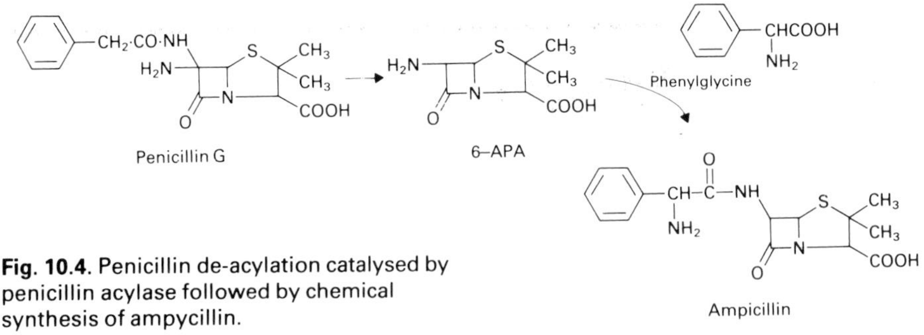

Fig. 10.4. Penicillin de-acylation catalysed by penicillin acylase followed by chemical synthesis of ampycillin.

Steroid modifications using enzymes are of great interest to the pharmaceutical industry because they are required to be stereospecific and very selective, i.e. only a single carbon atom must be modified if the molecule is to be active.

These are a few of the uses for enzymes in industry. In the future, we are likely to see more enzymes in use. In particular, there is great interest in the development of cellulases and lignin-degrading enzymes for use in recycling waste plant material, and also in microbial oxidases such as those involved in the oxidation of methane and ammonia.

Epilogue

In this brief introduction to microbiology we have tried to bring you a flavour of the form, function and utility of micro-organisms. Over the last hundred years the study of these most remarkably diverse organisms has come a long way, but there are clear indications that we still have a lot to learn. It is perhaps appropriate that the research carried out over the last century should have yielded much, and promises to yield yet more to the layman who has paid for it, albeit through indirect means. More importantly, microbiology is entering its most exciting phase as the understanding of biological processes goes hand in hand with their exploitation.

Additional Reading

More detailed reading lists are given in the later volumes of this series to which the reader is referred. In addition, the following more general references may be of value.

Brock T.D., Smith D.W. & Madigan M.T. (1984) *Biology of Micro-organisms*. London: Prentice Hall.

Stanier R.Y., Adelberg E.A. & Ingraham J.L. (1977) *General Microbiology*. London: Macmillan.

Atlas R.M. & Bartha R. (1981) *Microbial Ecology: Fundamentals and Applications*. Berkshire: Addison-Wesley Publishers.

Lynch J.M. & Poole N.J. (1979) *Microbial Ecology: A Conceptual Approach*. Oxford: Blackwell Scientific Publications.

Ingraham J.L., Maaloe O. & Neidhart F.C. (1984) *Growth of the Bacterial Cell*. Oxford: Blackwell Scientific Publications.

The Symposia of the Society for General Microbiology published yearly by the Cambridge University Press also have many articles of general interest.

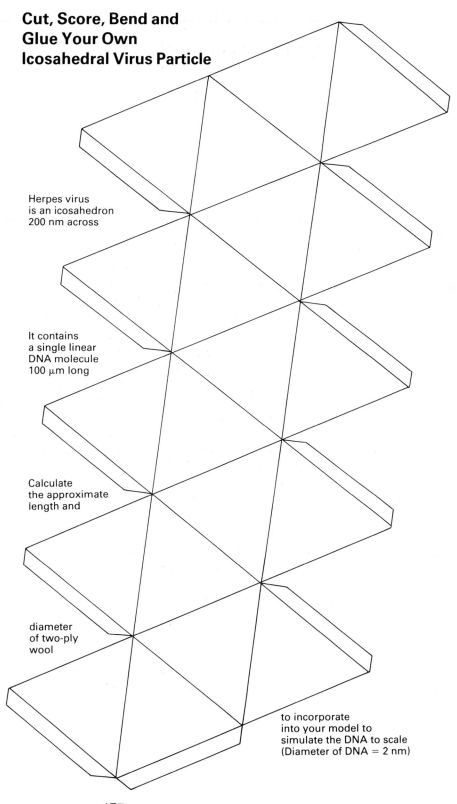

Cut, Score, Bend and Glue Your Own Icosahedral Virus Particle

Herpes virus is an icosahedron 200 nm across

It contains a single linear DNA molecule 100 μm long

Calculate the approximate length and

diameter of two-ply wool

to incorporate into your model to simulate the DNA to scale (Diameter of DNA = 2 nm)

Index

Introduction to Microbiology
Volume 1 of
Basic Microbiology
a series edited by
J.F. Wilkinson

Introduction to Microbiology

BASED ON THE ORIGINAL TEXT BY

J.F. WILKINSON MA, PhD
Professor of Microbiology
University of Edinburgh

THIRD EDITION REVISED AND EDITED BY

I.R. BOOTH BSc, PhD, G.W. GOODAY BSc, PhD

N.A.R. GOW BSc, PhD, W.A. HAMILTON PhD, FRSEd

& J.I. PROSSER BSc, PhD

All of the Department of Microbiology
University of Aberdeen

BLACKWELL SCIENTIFIC PUBLICATIONS

OXFORD LONDON EDINBURGH

BOSTON PALO ALTO MELBOURNE

Photoset by Enset (Photosetting),
Midsomer Norton, Bath, Avon
and printed and bound by
Billing & Sons Limited, Worcester.

DISTRIBUTORS

USA and Canada
    Blackwell Scientific Publications Inc
    PO Box 50009, Palo Alto
    California 94303

Australia
    Blackwell Scientific Publications
        (Australia) Pty Ltd
    107 Barry Street,
    Carlton, Victoria 3053

British Library
Cataloguing in Publication Data

Introduction to microbiology.—3rd ed.—
    (Basic microbiology; v. 1)
    1. Micro-organisms
    I. Booth, I.R.   II. Wilkinson, J.F.
    III. Series
    576   QR41.2

ISBN 0-632-00866-0